Galaxies: A Very Short Introduction

VERY SHORT INTRODUCTIONS are for anyone wanting a stimulating and accessible way in to a new subject. They are written by experts, and have been published in more than 25 languages worldwide.

The series began in 1995, and now represents a wide variety of topics in history, philosophy, religion, science, and the humanities. Over the next few years it will grow to a library of around 200 volumes – a Very Short Introduction to everything from ancient Egypt and Indian philosophy to conceptual art and cosmology.

Very Short Introductions available now:

Available soon:

For more information visit our website
www.oup.com.uk/vsi
www.oup.com/us

John Gribbin

GALAXIES
A Very Short Introduction

OXFORD
UNIVERSITY PRESS

OXFORD

UNIVERSITY PRESS

Great Clarendon Street, Oxford OX2 6DP

Oxford University Press is a department of the University of Oxford.
It furthers the University's objective of excellence in research, scholarship,
and education by publishing worldwide in

Oxford New York

Auckland Cape Town Dar es Salaam Hong Kong Karachi
Kuala Lumpur Madrid Melbourne Mexico City Nairobi
New Delhi Shanghai Taipei Toronto

With offices in

Argentina Austria Brazil Chile Czech Republic France Greece
Guatemala Hungary Italy Japan Poland Portugal Singapore
South Korea Switzerland Thailand Turkey Ukraine Vietnam

Oxford is a registered trade mark of Oxford University Press
in the UK and in certain other countries

Published in the United States
by Oxford University Press Inc., New York

British Library Cataloguing in Publication Data

Data available

Library of Congress Cataloging in Publication Data

Data available

ISBN 978-0-19-923434-9

1 3 5 7 9 10 8 6 4 2

Typeset by SPI Publisher Services, Pondicherry, India
Printed in Great Britain by
Ashford Colour Press Ltd, Gosport, Hampshire

For my brother, who suggested I write it.

Contents

Contents

List of illustrations

The publisher and the author apologize for any errors or omissions in the above list. If contacted they will be pleased to rectify these at the earliest opportunity.

Introduction

The scientific investigation of galaxies began only a little more than the 'three score years and ten' of a biblical lifetime ago, in the 1920s, when it was first established that many of the fuzzy blobs of light seen through telescopes are islands in space made up of vast numbers of stars, far beyond the boundaries of the Milky Way, our own island galaxy. Without telescopes, we would never have been able to explore the Universe beyond the Milky Way and investigate the nature of galaxies, but it had taken nearly four hundred years for telescopes to be developed to the point where the true nature of galaxies became clear.

As far as anyone knows, the first person to use a telescope to look at the night sky was Leonard Digges, an Oxford-educated mathematician and surveyor, who invented the theodolite some time around 1551. He kept his use of the telescope (essentially, a theodolite pointed upwards) secret, because of the value of the theodolite in his work, but he wrote one of the first popular books of what would now be called science in English. This included a description of the Ptolemaic Earth-centred model of the Universe. Leonard died in 1559, but his son, Thomas Digges, carried on where his father left off. Born in the 1540s, Thomas became a mathematician, and in 1571 arranged posthumous publication of a book by his father in which a telescope was described for the first time in print. Thomas Digges also made astronomical observations, and in 1576 published a revised and expanded version of his father's first book, which included the first printed description in English of the Sun-centred Copernican model of the Universe.

In that book, *Prognostication Everlasting*, the younger Digges said that the Universe is infinite, and included an illustration of the Sun, orbited by the planets, at the centre of an array of stars extending to infinity in all directions. Since we know Digges had at least one telescope, the natural inference to draw is that he had used one to look at the band of light across the sky known as the Milky Way, and had discovered that it is composed of uncountable numbers of individual stars.

The story of Leonard and Thomas Digges may come as a surprise, since the person usually credited both with making the first use of an astronomical telescope and with the discovery that the Milky Way is made of stars is Galileo Galilei, at the end of the first decade of the 17th century. In fact, the telescope was invented independently several times in north-west Europe, and news of the invention only reached Italy, from the Netherlands, in 1609. With only a description of the instrument to go on, Galileo built one of his own – the first of many – and among other things turned it on the heavens. His discoveries were published in a book titled *Sidereus Nuncius* (*The Starry Messenger*) published in 1610. This made him famous, and is the source of the popular myth that he was the first astronomer to use a telescope. But, like Thomas Digges before him, Galileo did indeed observe that the Milky Way is made up of a myriad of stars.

The next step towards an understanding of our place in the Universe was made by Thomas Wright, an English instrument maker and philosopher, in the middle of the 18th century. But, like that of the Digges, his contribution was almost forgotten. The Milky Way forms a band of light across the night sky, and in his book *An Original Theory or New Hypothesis of the Universe*, published in 1750, Wright suggested that it is made up of a slab of stars, which he likened to the shape of the grinding wheel of a mill. Even more impressively, he realized that the Sun is not at the centre of this disc-shaped slab of stars, but out to one side. He even suggested that the fuzzy blobs of light visible though telescopes, known as nebulae from their resemblance to clouds, might lie outside the Milky Way – although he did not make the leap of imagination that would have been required to suggest that the nebulae might be other star systems like the Milky Way itself. Immanuel Kant, another philosopher-scientist, picked up these ideas from Wright; he did take the extra step, and suggested that the nebulae might be 'island universes' like the Milky Way. The idea was not taken seriously.

As telescopes were improved, more and more nebulae were discovered and catalogued. One reason for the careful cataloguing was that astronomers of the late 18th and early 19th centuries were eager to find comets, and at first sight the fuzzy blob of a nebula looks like the fuzzy blob of a comet. So people like Charles Messier, in the 1780s, and William Herschel – who completed a catalogue in 1802 – identified the positions of nebulae in order that there should be no confusion. Herschel's catalogue included 2,500 nebulae, most of which we now know to be galaxies. Over the next 20 years, he tried to find out what the nebulae were made of, but even his largest telescope, with a mirror 48 inches (1.2 metres) in diameter, was unable to resolve the fuzzy patches of light into stars. He died in 1822 convinced that the nebulae really were diffuse clouds of material within the Milky Way.

The next observational step was made by William Parsons, the third earl of Rosse, who built an enormous telescope with a mirror 72 inches (1.8 metres) across in the 1840s. With this instrument, he found that many of the nebulae have a spiral structure, like the pattern of cream stirred into a cup of black coffee. Over the following decades, some nebulae were firmly identified as glowing clouds of gas within the Milky Way, and some were resolved into clusters of stars, on a much smaller scale than the Milky Way and associated with the Milky Way. But the spiral nebulae did not fit either category. The development of astronomical photography in the second half of the 19th century made it easier to study spiral nebulae, but the photographs were not good enough to reveal their true nature.

At the beginning of the 20th century, the vast majority of astronomers agreed that spiral nebulae were swirling clouds of material surrounding a star in the process of formation, like the cloud from which the Solar System was thought to have formed. But over the next two decades the island universes idea gained enough supporters to encourage the US National Academy of Sciences to host a debate on the subject, with Harlow Shapley, then of the Mount Wilson Observatory in California, speaking for the majority view against the island universe idea, and Heber Curtis, of California's Lick Observatory, speaking for it. Held on 26 April 1920, this became known to astronomers as 'The Great Debate'. Although it failed to resolve the issue, it marked the moment when the modern scientific investigation of galaxies began.

Chapter 1
The Great Debate

There were two aspects to the great astronomical debate of 26 April 1920: the size of the Milky Way Galaxy, and the nature of the spiral nebulae. In fact, it wasn't really a debate at all; the two speakers each made presentations 40 minutes long, and there was a general discussion afterwards. The theme of the meeting, held at what was then the US National Museum, and is now the Smithsonian Museum of Natural History, was 'The Scale of the Universe'. Shapley and Curtis had quite different views on what this meant, which they each elaborated on in a pair of scientific papers published the following year. In essence, Shapley thought that the Milky Way *was* the Universe, or at least the most important thing in the Universe, and was interested in the size of our own Galaxy; Curtis thought that the spiral nebulae were galaxies like our own, and was interested in the scale of things beyond the Milky Way.

The debate happened at the time it did because astronomers had recently developed techniques for measuring distances across the Milky Way. Distances to nearby stars can be measured using the same sort of surveying techniques with which Leonard Digges would have been familiar, starting with triangulation. If a nearby star is observed on nights six months apart, when the Earth is on opposite sides of its orbit around the Sun, the star will seem to shift slightly against the background of distant stars. This parallax effect is just like if you hold a finger up in front of your face and close each of your eyes in turn. The finger seems to move relative to the background, and the closer it is to your eyes the bigger this parallax effect is. The size of the stellar displacement and the diameter of the Earth's orbit (itself known from triangulation within the Solar System) are all you need to work out the distance to the star.

Unfortunately, most stars are far too far away for this effect to be measurable. Even the nearest star, Alpha Centauri, is so far away from the Sun that light takes 4.29 years to travel across the intervening space (so it is 4.29 light years away). By 1908, only about a hundred stellar distances had been measured in this way. Other geometrical techniques, based on the way with which stars in nearby clusters are seen to be moving together through space, make it possible to measure distances out to about 100 light years, or, in the units preferred by astronomers, about 30 parsecs (one parsec is almost exactly 3.25 light years). This was just enough for them to be able to calibrate the most important distance indicator in astronomy.

To put the importance of this new distance indicator in perspective we have only to look at the best estimate of the size of the Milky Way made in the early years of the 20th century. The Dutch astronomer Jacobus Kapteyn counted the number of stars visible on same-sized patches of sky in different directions from us, and included estimates of the distances to the stars, based on the techniques I have described and in part on how faint the stars seem from Earth. He inferred that the Milky Way was shaped rather like a discus, about 2,000 parsecs (2 kiloparsecs) thick in the middle, 10 kiloparsecs in diameter, with the Sun somewhere near the centre. This estimate is far too small, we now know, chiefly because there is a great deal of dust between the stars, which Kapteyn did not know, and this acts like a fog to limit how far we can see across the plane of the Milky Way – a phenomenon known as stellar extinction. Just as a traveller lost in fog seems to be alone at the centre of their own small world, so Kapteyn was lost in the fog of the Milky Way and seemed to be at the centre of his own small universe. Less than a hundred years ago, most astronomers thought that this discus of stars essentially represented the entire Universe.

Things began to change in the second decade of the 20th century. Henrietta Swan Leavitt, working at the Harvard College Observatory, discovered that a certain family of stars, known as Cepheids, vary in brightness in a way which makes it possible to use them as distance indicators. Each Cepheid brightens and dims in a regular way, repeating the cycle exactly again and again. Some run through the cycle in less than a day; others take as long as a hundred days. Polaris, the northern pole star, is a Cepheid with a period close to four days, although the brightness changes in this case are too small to be detected with the naked eye. Leavitt's great discovery was that brighter Cepheids take longer to run through their cycles than fainter Cepheids. Even better, there is an exact relationship between the period of a Cepheid and its brightness. A Cepheid that takes five days to run through its cycle, for example, is ten times brighter than one which takes eleven hours to run through its cycle.

Leavitt made this discovery by studying the light from hundreds of stars in a nebula called the Small Magellanic Cloud (SMC), a star system associated with the Milky Way. She didn't know how far away the SMC was, but that didn't matter because all the stars in it are at essentially the same distance from us. So their relative brightnesses could be compared without having to worry if one star looked fainter than another simply because it was farther away. In 1913, the Dane Ejnar Hertzprung measured the distances to 13 nearby Cepheids using geometrical techniques, and used observations of these stars combined with Leavitt's data to work out the true brightness of a hypothetical standard Cepheid with a period of exactly one day. Armed with this calibration, it was possible to measure the distance to any other Cepheid by working out its true brightness from Hertzsprung's calibration and its period, then comparing this with how faint the star appeared on the sky – the fainter it was, the farther away it must be, in a precisely calculable fashion. Among other things, this calibration of the Cepheid distance scale meant that the SMC is at least 10 kiloparsecs (kpc) away. Hertzsprung's estimate has since been revised in the light of better observations and an understanding of stellar extinction, but in 1913 it marked a dramatic increase in scale from Kapteyn's estimate of the size of the entire Milky Way (the entire Universe!) to suggest that the SMC was so far away.

It was Harlow Shapley who used the Cepheid technique to map the size and shape of the Milky Way Galaxy itself, after having carried out his own calibration of the brightnesses of these variable stars. This was at the heart of his contribution to the Great Debate.

The key to Shapley's survey of the Milky Way was that he was able to use variable stars to measure distances to star systems known as globular clusters. As their name suggests, globular clusters are spherical star systems. They may contain hundreds of thousands of individual stars, and at the heart of such a cluster there may be as many as a thousand stars packed into a single cubic parsec of space – very different from our region of the Galaxy, where there is *no* star as close as 1 parsec to the Sun. Globular clusters are seen above and below the plane of the Milky Way. By measuring the distances to them, Shapley found that they are distributed in a spherical volume of space centred on a point in the direction of the constellation Sagittarius but thousands of parsecs away from us, in the middle of the band of light known as the Milky Way. The implication is that this point marks the centre of the Milky Way Galaxy, and that the Solar System is, therefore, located far out towards the edge of the Galaxy. By 1920, Shapley had come up with an estimate that our Galaxy is about 300,000 light years (nearly 100 kpc) across, with the Sun about 60,000 light

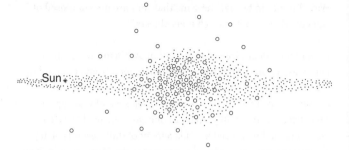

1. The distribution of globular clusters (circles) on one side of the sky implies that the Sun is far from the centre of the Milky Way

years (nearly 20 kpc) out from the centre. As he put it at the Washington meeting:

> One consequence of the cluster theory of the galactic system is that the sun is found to be very distant from the center of the Galaxy. It appears that we are near the center of a large local cluster or cloud of stars, but that cloud is at least 60,000 light years from the galactic center.

On this picture, it seemed to Shapley and like-minded astronomers that the spiral nebulae could not be other galaxies like the Milky Way. Their reasoning was simple. The apparent (angular) size of an object on the sky depends on its actual linear size and its distance from us, in exactly the same way that a real cow standing far away on the other side of a field looks the same size as a child's toy cow held up in your hand. If the spiral nebulae were also about 300,000 light years across, their tiny angular sizes on the sky would place them at distances of many millions of light years, which just seemed too big to take seriously. Instead, Shapley argued that the spiral nebulae were either star-forming systems within the Milky Way Galaxy, or at most small satellites of the Milky Way – islands compared with the continent of the Milky

Way. 'I prefer to believe', he said, 'that they are not composed of stars at all, but are truly nebulous objects.'

He had one other piece of ammunition. Adriaan van Maanen, a Dutch astronomer who happened to be a good friend of Shapley, claimed that he had measured the rotation of several spiral nebulae, by comparing photographs taken several years apart. The effect was incredibly small. In one case, the nebula M101, he said he had measured a displacement of 0.02 seconds of arc, about 0.001 per cent of the angular size of the Moon as seen from Earth. Any such rotation can be converted into a linear speed corresponding to the distance of any part of the nebula from the centre of rotation. This, of course, depends on the actual size of the object that is rotating. If the spirals were the same size as the Milky Way, van Maanen's measurements would have implied speeds comparable to, or even faster than, the speed of light. If he was right, the spirals had to be small objects, relatively close to us. Most astronomers found it hard to believe that van Maanen could actually be making such precise measurements. Later studies showed that van Maanen had made a mistake – nobody quite knows how – but at the time of the Great Debate it was a matter of faith whether you believed his data or not; and Shapley trusted his friend. In his paper published in 1921, Shapley emphasized that van Maanen's results 'appear fatal' to the island universe idea. 'Bright spirals cannot reasonably be the excessively distant objects required by the theory.'

Curtis didn't trust van Maanen's results, and he also didn't trust the still-new Cepheid distance scale. At the Washington meeting, he gave a summary of various earlier estimates of the size of the Galaxy, including, rather cheekily, an estimate for the diameter of just 20,000 light years made by Shapley in 1915. He concluded that 'a maximum galactic diameter of 30,000 light years will be assumed as representing sufficiently well the older view; it is perhaps too large'. This estimate was exactly one tenth of the size Shapley suggested in 1920. Curtis also said that the Sun is located 'fairly close to' the centre of the Galaxy, but not exactly at the centre. But all of this was, to him, a minor matter which he mentioned briefly before going on to discuss the aspect of the story that really interested him, the nature of spiral nebulae and their distances from us.

There were two key facts that Curtis used in his argument that the spiral nebulae are galaxies like our own at great distances from us. The first was the discovery, made by Vesto Slipher, of the Lowell Observatory, that by far the majority of spiral nebulae seemed to be receding from us at very high velocities. The discovery was made by measuring the extent to which lines in the spectra of these nebulae are displaced towards the red end of the spectrum, compared with lines in the light from nearby stars and hot objects on Earth.

Light from any hot object, including the Sun and stars, can be spread out using a prism to make a rainbow pattern, or spectrum. Each chemical element – hydrogen, carbon, and so on – produces a characteristic pattern of bright lines in the spectrum, as distinctive as the bar codes on supermarket goods. When the object is moving away from us, the whole pattern of lines is displaced towards the red end of the spectrum, by an amount which depends on how fast the object is receding; this is the famous redshift. Similarly, when an object is moving towards us, the pattern of lines is shifted towards the blue end of the spectrum – a blueshift. Stars moving around in the Galaxy show both redshifts and blueshifts, corresponding to velocities relative to us of anything from zero to a few tens of kilometres per second.

In the second decade of the twentieth century, measuring the positions of lines in the faint spectra of light from spiral nebulae was pushing photographic techniques to the limit. It was only in 1912 that Slipher was able to obtain such spectrographs of the Andromeda Nebula, also known as M31, which we now know to be the nearest spiral to the Milky Way. He found a shift towards the blue end of the spectrum, indicating that the nebula is rushing towards us at 300 kilometres per second. This was by far the highest such speed measured up to that time. By 1914, Slipher had similar spectrographs for 15 spirals. Only two, including M31, showed a blueshift. The other 13 all showed redshifts, two of them corresponding to velocities of recession of more than a 1,000 kilometres per second. By 1917, he had 21 redshifts, but still only two blueshifts – even today, there are still only two blueshifts. Whatever the nature of spiral nebulae, Slipher's measured velocities implied that they could not be part of the Milky Way; they were simply moving too fast to be gravitationally bound to our Galaxy. Although in 1920 nobody could explain the cause of these large recession velocities, Curtis saw this as evidence that the spiral nebulae had nothing to do with our Milky Way, but were 'island universes' in their own right.

The other main plank in his platform concerned observations of stars that suddenly flared up in bright outbursts. Such stars are known as novae, from the Latin word for 'new', because when they were first observed they literally seemed to be new stars, shining brightly where no star had been noticed before. It is now clear, however, that all novae are outbursts from stars which had previously been leading a quiet life and were too faint to be seen. They are a natural, but fairly rare, stellar phenomenon.

In 1920, Curtis pointed out that 'within the past few years some twenty-five novae have been discovered in spiral nebulae, sixteen of these in the Nebula of Andromeda, as against about thirty in historical times within our own galaxy'. The sheer number of novae seen in the Andromeda Nebula meant that the nebula must be made up of a huge number of stars, assuming that a star in Andromeda was no more likely to become a nova than a star in the Milky Way, and roughly speaking the apparent brightness (faintness) of the novae seen in various spirals was about what you would expect if they were actually as bright as novae in the Milky

2. A classic example of a disc galaxy. This is NGC 4414, viewed by the Wide Field Planetary Camera 2 (WFPC2) on the Hubble Space Telescope

Way, but as remote as the distances implied if the spiral nebulae were the same size as Curtis's estimate of the size of the Milky Way.

There was one fly in the ointment. In 1885, in the very decade that the Andromeda Nebula was identified as a spiral, a bright star flared up in it. The apparent brightness of this nova was about the same as the apparent brightness of a typical nova in the Milky Way. This meant that either the nebula really was part of the Milky Way, or, if the nebula was as far away as Curtis thought, that this was some kind of super-powerful nova, as bright as a billion Suns put together and far brighter than any nova observed in the Milky Way in the 19th century. This was a difficulty for Curtis, which he essentially circumvented by suggesting that there might be two kinds of nova, one much brighter than the other. This seemed like a fudge to his audience at the time; but we now know that there really are stellar outbursts that bright. They are called supernovae, and they can briefly shine as brightly as a hundred billion Suns – as bright, indeed, as all the other stars in a galaxy put together.

As Curtis summed the argument up:

> The new stars observed in the spirals seem a natural consequence of their nature as galaxies. Correlations between the new stars in spirals and those in our Galaxy indicate a distance ranging from perhaps 500,000 light years in the case of the Nebula of Andromeda, to 10,000,000, or more, light years for the more remote spirals ... At such distances, these island universes would be of the order of size of our own Galaxy of stars.

In the paper published in 1921, he went further:

> the spirals, as external galaxies, indicate to us a greater universe into which we may penetrate to distances of ten-million to a hundred-million light-years.

In so far as there was a debate on the scale of the Universe in Washington on 26 April 1920, nobody won. Both participants were sure they had come out on top – a sure sign that neither of them had – but both were right on some points and wrong on others. Most importantly, Shapley was right to trust the Cepheid distance scale, even though it hadn't quite been perfected at the time, and Curtis was right that the spiral nebulae are other galaxies. Shapley was also right in placing the Sun far out from the centre of the Milky Way. As for the size of the Milky Way, the best current estimates give a diameter of about 100,000 light years, three times bigger than Curtis's estimate and one third the size of

Shapley's estimate, so by that reckoning they were equally wrong. This does indeed make the Milky Way an average spiral – just how average, I shall discuss in Chapter 4. Although the Great Debate was inconclusive, the key issues it raised were resolved before the end of the 1920s, largely thanks to the work of one man, Edwin Hubble.

Chapter 2
Stepping stones to the Universe

The main reason why the study of galaxies took off in the 1920s was the invention of bigger telescopes and improved photographic techniques, which made it possible to obtain more detailed images (and spectra) of faint and distant objects. Spectrophotography was vital to the discovery of redshifts in the light from spiral nebulae, and photography itself was a key element in the discovery of the Cepheid period–brightness relation. In 1918, a telescope with a 100-inch (2.5-metre) diameter mirror became operational on Mount Wilson in California; it would be the most powerful telescope in the world for almost three decades, and was used by Edwin Hubble to measure the distances to galaxies in a series of steps out across the Universe.

Hubble cut his teeth as a research astronomer as a Ph.D. student at the Yerkes Observatory (part of the University of Chicago) between 1914 and 1917. His research project there was to obtain photographs of faint nebulae using a 40-inch (1-metre) refracting telescope. This was one of the best telescopes in the world at the time, and the largest refractor ever built. By and large, for telescopes the same size, refractors, which use lenses, are more powerful than reflectors, which use mirrors; but reflectors can be made bigger because their mirrors can be supported from behind without blocking out any light. This observing programme led Hubble to study the nature of nebulae and to a classification of nebulae based on their appearance. It also convinced him, by 1917, that the great spirals, in particular, must lie beyond the Milky Way.

The development of these ideas was delayed because as soon as Hubble had completed his Ph.D. he volunteered to serve in the US Army, following the United States's entry into the First World War in April 1917. He served in France and reached the rank of major, but never saw action. It wasn't until September 1919 that Hubble eventually joined the staff at Mount Wilson Observatory, where he was one of the first people to use the new 100-inch telescope. He also took the opportunity to develop the ideas from his Ph.D. thesis into a full classification scheme which he completed in 1923. Hubble always used the term nebulae for the objects he was describing, but he was convinced that they lay outside the Milky Way; as he was soon proved right, in line with modern usage I shall call them galaxies. The most important thing to emerge from Hubble's early work is that there are indeed different kinds of galaxy, and the giant spirals are simply the most obvious of these objects.

Apart from a relatively small number of relatively small, irregularly shaped galaxies like the Small Magellanic Cloud (and its bigger counterpart the Large Magellanic Cloud), all galaxies can be defined according to their shape. The term elliptical galaxy is used for those which appear to be anything from spherical to the shape of an elongated lens, but have no obvious internal structure. Spirals may have more tightly wound or more open spiral structures, and in all cases there are examples in which the spiral arms start at the centre of the galaxy, and examples in which the spiral arms seem to be connected to the ends of a bar of stars across the centre of the galaxy. Hubble thought that there was an evolutionary sequence in which an open spiral of either kind gradually became more and more tightly wound, as a result of rotation, and ended up as an elliptical. He was completely wrong, but this does not affect his classification scheme based on the appearance of galaxies. We now know that the largest galaxies in the Universe are giant ellipticals, but some ellipticals are smaller than some spirals. We also know that some of the galaxies originally regarded as 'spirals' are disc-shaped systems of stars with no discernible spiral arms at all! For this reason, it is better to use the term 'disc galaxy', which includes the ones with spiral arms; but even today many astronomers refer to 'spirals' when they are talking about essentially featureless disc galaxies.

Hubble's career at Mount Wilson overlapped briefly with that of Harlow Shapley, who left to take up a post at Harvard in March 1921. By the time Hubble began using the 100-inch telescope to try to prove that the nebulae he had been studying were other galaxies, the more senior astronomer was no longer around to object. With ever-improving observations, the island universe idea was, in any case, beginning to gain support in the early 1920s. A Danish astronomer, Knut Lundmark, who visited both the Lick Observatory and the Mount Wilson Observatory at that time, obtained photographic images of a nebula (galaxy) known as M33 which were good enough to convince him, although not Shapley, that the granulated appearance of the image showed that the nebula was made of stars. In 1923, several variable stars were discovered in the nebula NGC 6822, but it took a year before they could be identified as Cepheids, and by then Hubble had already made the breakthrough discovery of Cepheids in M31, the Andromeda Nebula.

He wasn't actually looking for Cepheids. With his classification scheme completed, in the autumn of 1923 Hubble followed up one of the main lines of Curtis's argument by starting a series of photographic observations with the 100-inch telescope, aimed at discovering novae in one of the spiral arms of M31. Almost immediately, in the first week of October that year, he found three bright spots of light which looked like novae on the photographic plates. Because the 100-inch telescope had been operating for several years, there was already an archive of photographs which included observations of the same part of M31, obtained by several observers, including Shapley and Milton Humason, who was to become Hubble's closest collaborator in the years that followed. These plates showed that one of the three bright spots that Hubble had tentatively identified as novae was, in fact, a Cepheid, with a period of a little more than 31 days. Using Shapley's calibration of the Cepheid distance scale, this immediately gave a distance of nearly a million light years (300 kpc), three times bigger than even Shapley's estimate of the size of the Milky Way Galaxy. The whole distance scale was later revised, partly because of the problems caused by interstellar extinction, and we now know that M31 is actually about 700 kpc away, roughly equivalent to about 20 times the diameter of the Milky Way. But what mattered in 1923 was that at a stroke, with almost his first observations of the nebula, Hubble had shown that it is indeed a galaxy more or less like our own, located far outside the Milky Way.

Over the following months, Hubble found one more Cepheid and nine novae in M31, all giving him roughly the same distance estimate, and other Cepheids and novae in other nebulae. He put everything together in a paper presented to a joint meeting of the American Astronomical Society and the American Association for the Advancement of Science held in Washington, DC, on 1 January 1925. Hubble was not present at the meeting, where the paper was read on his behalf by Henry Norris Russell. But there was no need for personal advocacy. The consensus of the meeting was that the nature of the nebulae had at last been determined, and that the Milky Way Galaxy is just one island in a much bigger Universe. Even before that meeting, Hubble had written to Shapley to tell him about the discoveries. The astronomer Cecilia Payne-Gaposchkin, who had started her Ph.D. studies under Shapley's supervision in 1923, happened to be in his office when he read the letter. 'Here', he said as he offered it to her, 'is the letter that destroyed my universe.' The Great Debate was over. It may have been some consolation for Shapley that Hubble's successful

3. The dome of the 100-inch Hooker Telescope on Mount Wilson, used by Edwin Hubble to measure distances to galaxies

use of the Cepheid technique lent weight to Shapley's model of the Milky Way, and in particular to the displacement of the Sun from the centre of our Galaxy.

But if Shapley's universe had been destroyed, what was the new universe – Hubble's universe – like? The Universe is so big that even with the 100-inch telescope Hubble was only able to obtain images of Cepheids in what turned out to be very nearby galaxies. Observers working with lesser telescopes were even more handicapped. Fascinated – almost obsessed – with the idea of measuring the scale of the Universe, Hubble had to find other ways to measure the distances to galaxies beyond the range of the Cepheid technique, and he set about this task in the middle of the 1920s.

Hubble put together a series of stepping stones which observers could use to reach farther and farther out into the Universe. Cepheids are just bright enough to give the distances to a few of the nearest galaxies, only a few dozen before the advent of the space telescope named after Hubble and launched in 1990, but novae are a little brighter than Cepheids and can be seen farther away. Once the distance to M31 had been determined from Cepheids, Hubble was able to use this to calibrate the brightness of novae seen in that galaxy, and then, making the assumption that all novae have the same intrinsic brightness, to use observations of novae to measure distances to galaxies that are a little farther away. With the ability of the 100-inch and its successors to resolve individual stars in nearby galaxies, other techniques became feasible. The brightest individual stars in galaxies are also much brighter than Cepheids, and could be used as distance indicators in the same way, this time making the assumption that the brightest stars in any galaxy will be about as bright as the brightest stars in any other galaxy, since there must be some upper limit to how bright a star can be. He could also identify globular clusters in other galaxies, and guess that the brightest globulars in each galaxy must have roughly the same intrinsic brightness as each other. Supernovae, once they were understood, were later added to the chain in the same way.

More rough and ready estimates were based on the brightnesses of whole galaxies, and on their apparent (angular) sizes on the sky. If every spiral galaxy was exactly as bright as M31 and each the same size as M31, it would be easy to measure their distances by comparing their observed properties with the properties of M31. Unfortunately, this is far from being the case, and Hubble knew it; but for want of anything better, he tried to compare the observed properties of galaxies that looked much the same as one another in order to get at least some guide to their distances.

None of these techniques is perfect, but wherever possible Hubble applied as many of the techniques as he could to each individual galaxy, hoping that the errors and uncertainties would average out. This all took time, but in 1926 Hubble had begun to build up a picture of the distribution of galaxies around the Milky Way Galaxy. There was just enough data for him to contemplate taking

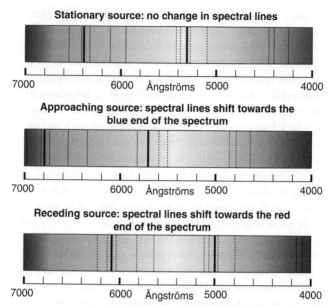

Stationary source: no change in spectral lines

7000   6000   Ångströms   5000   4000

Approaching source: spectral lines shift towards the blue end of the spectrum

7000   6000   Ångströms   5000   4000

Receding source: spectral lines shift towards the red end of the spectrum

7000   6000   Ångströms   5000   4000

4. The velocity and direction of travel of the star in relation to the observer determine the amount by which the banding in the spectrum shifts. When a radiating body is moving away from the observer, the waves emitted become 'stretched', the wavelength lengthens, and the spectral lines shift to the red end of the spectrum. If the body is approaching, the wavelength is compressed, and the lines shift towards the blue end of the spectrum. Redshift can be used to calculate an object's recession velocity

a great leap outward by following up a hint already present in the redshift data obtained by Vesto Slipher and a few other people.

By 1925, analyses of the light from what were now known to be other galaxies had revealed 39 redshifts and still the same two blueshifts. In fact, Slipher had been the first person to measure all but four of these redshifts, but he soon reached the limit of what was possible with the telescope he was using at the Lowell Observatory, a 24-inch (60-cm) refractor, ending up with 43 redshifts. There was a hint – barely – from these data that larger redshifts were associated with more distant galaxies. Several people had noticed this, but Hubble, by now an established astronomer with access to the best telescope in the world, was the man in the right place at the right time to try to prove that this was the case. His motivation was to find out whether there was a precise relationship between redshifts and distances that he could use as the final step in his chain, making it possible to measure distances across the Universe simply by measuring redshifts.

In 1926 Hubble deliberately set out to find a link between redshifts and distances for galaxies. He already had many distances, and would determine more over the years that followed, but the 100-inch had never been used for redshift work, and he needed a colleague able and willing to set the telescope up for this difficult work, and then carry out the painstaking measurements. He chose Milton Humason, a superb observer but clearly junior to Hubble, so that it would be obvious to the outside world who was the team leader. After all the hard work of adapting the 100-inch to its new role, Humason deliberately chose for his first redshift measurement a galaxy too faint to have been studied in this way by Slipher. He obtained a redshift corresponding to a velocity of about 3,000 km per second, more than twice as large as any redshift measured by Slipher. The Hubble–Humason partnership was up and running.

By 1929, Hubble was convinced that he had found the relationship between redshift and distance. Not only that, it was the simplest relationship he could have hoped to find – redshift is proportional to distance, or, putting it the way round that mattered to Hubble, distance is proportional to redshift. A galaxy with a redshift twice as large as that of another galaxy is simply twice as far away as the nearer galaxy. The first results of the collaboration, published in 1929, gave data for just 24 galaxies which had both known redshifts and known distances, from which Hubble calculated that the constant of proportionality in the redshift–distance relation was 525 km per second per Megaparsec. That is, a galaxy with a redshift corresponding to a velocity of 525 km per second would be one million parsecs (3.25 million light years) away, and so on. The choice of this particular number looked as much like wishful thinking as anything else, because the limited amount of data was not really good enough to justify the precision of the quoted number. But in 1931 Hubble and Humason together published a paper updating these results with a further 50 redshifts, going out to a distance equivalent to a velocity of 20,000 km per second, and fitting the number Hubble had obtained three years earlier much more closely. Clearly, he had already had some of these data in 1929, but had chosen, for whatever reason, not to publish them at the time.

Hubble neither knew nor cared why the redshift–distance relation existed. He didn't even claim that it meant that other galaxies are moving away from us. Although redshifts are conventionally quoted in units of km per second, there are other ways than motion through space known to produce them (for example, a strong gravitational field) and Hubble was careful to consider that processes unknown in the 1930s might be at work. In his book *The Realm of the Nebulae*, he wrote:

> Red-shifts may be expressed on a scale of velocities as a matter of convenience. They behave as velocity-shifts behave and they are very simply represented on the same familiar scale, *regardless of the ultimate interpretation*. The term 'apparent velocity' may be used in carefully considered statements, and the adjective always implied where it is omitted in general usage. (Emphasis added.)

Whatever the origin of the redshift–distance relation, it did indeed prove the ultimate tool for measuring the scale of the Universe,

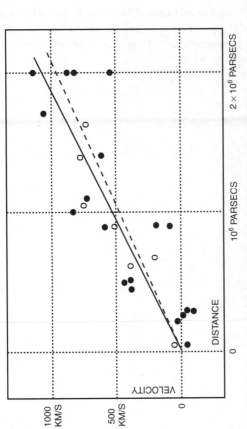

5. Hubble's original 'redshift–distance relation' was based on a rather optimistic interpretation of the data published in 1929; by 1931, his work with Humason made a much more convincing case

and the constant of proportionality became known as Hubble's Constant, or H. Since 1931, the aim of all other measurements of distances beyond the Milky Way has simply been to calibrate the Hubble Constant. But before looking at what all this implies for the understanding of galaxies and their place in the Universe at large, it seems appropriate to sum up the present understanding of our own home in space, the Milky Way, an ordinary spiral galaxy.

Chapter 3
Our island

Since the 1920s our understanding of the Milky Way Galaxy has increased dramatically, largely because of the continuing development of observing techniques and technology. As well as having larger and better telescopes to observe in visible light (including the Hubble Space Telescope), we have data obtained by radio telescopes, in the infrared part of the spectrum and from X-ray detectors and other instruments carried into space on satellites. Sensitive electronic detectors are able to obtain much more information about faint sources than is available from photographs or from the kind of spectroscopic instruments available to Hubble and his contemporaries, and the power of modern computers makes it much easier than it was in his day to compare theoretical predictions with observations.

The most profound discovery about the Milky Way made since the 1920s is that all of the bright stars make up only a small fraction of the total amount of mass in our Galaxy. From the way that the whole system rotates, it is clear that the bright disk is held in the gravitational grip of a roughly spherical halo of dark matter which has about seven times as much mass as everything that Hubble would have thought of as the Galaxy put together. This has profound implications for our understanding of the Universe at large, since the same ratio of ordinary matter to dark matter seems to apply across the Universe. These cosmological implications are

discussed by Peter Coles in *Cosmology: A Very Short Introduction*.
The most important point, apart from the existence of the dark
matter, is that it is not simply cold gas or dust. It is not made of
the kind of particles – atoms and so on – that the Sun and stars,
and ourselves are made of, but is something else entirely. Since
nobody knows exactly what it is, it is simply referred to as Cold
Dark Matter, or CDM.

Our Sun is a typical star. Some contain more mass, some less,
but they all work in the same way, converting light elements (in
particular hydrogen) into heavier elements (in particular, helium)
in their interiors by nuclear fusion, releasing the energy that keeps
the stars shining. Overall, it is estimated that there are several
(at least three) hundred million stars in the Milky Way, spread
across a disc about 28 kiloparsecs (just over 90 thousand light
years) across. There is some uncertainty about the exact size (it
is difficult to measure the size of a forest from inside it), so these
numbers are often rounded off to 30 kpc and 100,000 light years.
There is a bulge of stars in the centre of the disc, which would
give it the appearance, if viewed edge-on from outside, of two
fried eggs stuck back to back. The whole disk is surrounded by the
spherical halo of old stars and globular clusters, which contain the
oldest stars in the Galaxy. Nearly 150 globular clusters are known,
and there must be another 50 or so that we cannot see because the
bright band of light of the Milky Way is between us and them.

Astronomers can study the way stars move through space using
the Doppler effect. This shifts the lines in the spectrum of a star
towards the red end of the spectrum if it is moving away, and
towards the blue if it is moving towards us. The size of the effect
reveals the star's velocity. This is exactly equivalent to the way the
sound emitted from a moving object – for example, the siren on an
ambulance – is deepened if the vehicle is moving away but raised
in pitch if it is moving towards you. Christian Doppler predicted
the effect in 1842, then measured it using trumpeters blowing
a steady note as they moved past on a train. Superficially, the

6. The starburst galaxy M82. This is a composite image combining data from WFPC2 and the 3.5-metre telescope on Kitt Peak in the United States

effect resembles the redshift seen in the light of galaxies; but the cosmological redshift is not caused by motion through space and is not a Doppler effect.

The Sun is about two-thirds of the way out (a bit less than 10 kpc) from the centre of the Milky Way to the edge of the visible disc. Like other stars in the disc, it moves around the centre of the Galaxy, at a speed of about 250 km per second in a roughly circular orbit, and takes a little less than 250 million years to complete one circuit. The ages of stars can be determined by comparing their overall appearance (especially their colour and brightness) with theoretical models of how they change as they consume their nuclear fuel; in the case of the Sun this is confirmed by using measurements of the radioactivity in rocks and meteorites to infer the age of the Solar System. The Sun and Solar System have been around for about 4.5 billion years, long enough to have completed about 20 orbits of the galactic centre. Since the first humans, modern *Homo sapiens*, appeared on Earth, the Solar System has completed just under one-thousandth of its present circuit. The oldest stars are a little over 13 billion years old, three times the age of the Sun.

Outside the central bulge, the disc of the Galaxy is only about 300 kpc (roughly 1,000 light years) thick. The Solar System is only 6 or 7 parsecs above the centre of the plane of the disc. Viewed from above, the Galaxy's resemblance to a fried egg would be spoiled by the bar, 8 or 9 kpc long, across the middle of the bulge, but it would be possible to pick out four quite tightly wound spiral arms twining outward from the centre. As in other disk galaxies, the spiral arms are bright because they contain many hot young stars in the first flush of youth. These stars are big as well as bright. The bigger (more massive) a star is, the more intensely it has to burn its nuclear fuel to hold itself up against the pull of gravity, and the more quickly it uses up its fuel. Spiral arms are the sites of star formation. Smaller, long-lived stars, like the Sun, are also formed in spiral arms, but do not burn so brightly. The Solar System is at present in a lesser spur of stars known as the Orion Arm, or simply the Local Arm, which forms a kind of bridge between two of the major arms. Shapley was right in thinking that we are in a large local concentration of stars.

The young stars that are found chiefly in the spiral arms and the plane of the Milky Way (and in the discs of other galaxies) are known as Population I. The Sun is a Population I star. They contain recycled material from previous generations of stars,

7. **Star-forming region in Orion, imaged in the infrared by the Spitzer Space Telescope**

including the heavy elements from which planets like the Earth are composed. Older stars which are found in the halo of the Galaxy, in globular clusters and in the bulge, are known as Population II. These old stars tend to be redder than Population I stars. They formed long ago when the Galaxy was young, and are chiefly composed of the primordial hydrogen and helium that emerged from the Big Bang in which the Universe was born. The heavier elements in Population I stars and in ourselves were made inside previous generations of stars. Elliptical galaxies are largely made up of Population II stars.

If the spiral pattern seen in a galaxy like the Milky Way was not maintained in some way, it would soon get smeared out, within about a billion years, as the stars moved around the Galaxy in their orbits. It persists because it is a wave of star formation maintained by clouds of gas and dust moving around the Galaxy in their own orbits and being squeezed as they cross the spiral arms. The young stars are simply the most visible feature of a shock wave travelling around the Galaxy, similar to the shock wave of a sonic boom.

An analogy that is often made is with the kind of traffic jam that occurs on a busy motorway when there is a large, slow-moving vehicle occupying the inside lane. As the faster traffic comes up behind the large vehicle it is squeezed into the outer lanes and makes a moving traffic jam which disperses on the other side of the large vehicle. The traffic jam moves along the motorway at a steady speed, but it is constantly changing as new cars join the back and others leave the front. In the same way, a spiral arm moves around the Galaxy at a constant speed, but new clouds of gas and dust are constantly joining it, being squeezed, and then going on their way. Some of these clouds get squeezed sufficiently to trigger star formation, in a self-sustaining process.

But although it is self-sustaining, this is not a very efficient process. If it were, then by now the Milky Way would have formed all of the gas and dust it contains into stars. In fact, only a few times as much material as there is in the Sun (a few solar masses of material) is converted into new stars each year in our Galaxy. This roughly balances the amount of material thrown back out into space by old stars when they die, so the processes of star birth, life, and death are able to continue for many billions of years in a disc galaxy. This also implies that very many stars must have been born in a short space of time when the Milky Way formed, before it settled down. Such spectacular events, known as starbursts, are indeed seen in other galaxies.

It is difficult for a cloud of gas and dust to collapse to form a star (or several stars) for two reasons. First, all clouds are rotating, if only slightly, and as they contract they will spin faster, resisting the pull of gravity. They have to break up in such a way that their angular momentum is dissipated in some way. Second, a collapsing cloud will get hot, as gravitational energy is released, and unless this heat can be radiated away it will prevent any further collapse. The angular momentum problem is solved by clouds breaking up into several stars, so that the angular momentum of the cloud is converted into the angular momentum of the stars orbiting one another. On average, out of every 100 newly born star systems, 60 are binaries and 40 are triples. Solitary stars like the Sun are later ejected from triple systems formed in this way. The heat problem is solved because the clouds contain molecules such as carbon monoxide, which warm up and radiate the heat away in the infrared part of the spectrum. But star formation is still a difficult process – the wonder is that there are any stars at all.

Star formation begins in large complexes of gas, perhaps a thousand parsecs across and containing ten million solar masses of material, within which an individual cloud may be a few tens of parsecs across and contain a few hundred thousand solar masses of material. The initial squeeze to cause the collapse of a cloud most probably comes from the explosion of a massive star, a supernova. Turbulence within the collapsing cloud leads to the formation of cores about a fifth of a light year across, containing about 70 per cent as much mass as our Sun. But only a few per cent of the mass of the whole cloud gets converted into cores in this way. When a star forms, it starts out as an even smaller inner core, with only one thousandth of the mass of the Sun, reaching the density necessary to turn itself into a star. The rest of the mass of the star is added as material from the surrounding cloud close enough to be pulled in by gravity falls on to the core, so the eventual mass of the star depends on how much material there is nearby. Once the stars begin to shine, the radiation from them blows away the rest of the surrounding material.

The whole process is over very quickly. A cloud collapses to make stars and the hot young stars blow away the leftover material to leave behind a cluster of stars, all within about ten million years. The late stages of this process can be seen in the nearby Orion Nebula. But some of the young stars in some clusters will be much more massive than the Sun, and will use up their nuclear fuel very quickly. These are the stars which end their lives by exploding as supernovae, sending shock waves out through the interstellar material and triggering the collapse of other clouds of gas and dust. This seems to be a self-sustaining process which keeps a galaxy like the Milky Way in a steady state with the aid of negative feedback. If a larger than average number of stars form in one generation or one location, the energy from them will disperse the gas and dust over a wide region, reducing the number of stars in the next generation; but if only a few stars form, there will be plenty of gas and dust left over to make more stars next time the cloud is squeezed. The natural tendency is for the process to shift back towards the average. And because the kind of stars that form supernovae burn out in only a few million years (compare that with the 4.5 *billion* years that the Sun has been around so far) all of this activity takes place within the vicinity of the spiral arms, helping to maintain the spiral pattern.

The central region of our Galaxy, around which the whole spiral pattern rotates, is more than just the mathematical centre of the disc. There is a black hole containing 2.5 million times as much mass as our Sun at the centre of the Milky Way, and as we shall see in Chapter 7, such black holes hold the key to the existence of galaxies.

Most popular accounts of black holes concentrate on much smaller objects, with masses only a few times that of our Sun. Such objects form if a star at the end of its life has more than about three times the mass of the Sun today. Such a stellar cinder, no longer generating heat in its interior because all of its fuel is exhausted, cannot hold itself up under its own weight, and collapses, shrinking (according to the general theory of relativity) into a point of zero volume, called a singularity. Atoms and the particles they are made of, protons, neutrons, and electrons, are crushed out of existence in the process. Almost certainly, the general theory of relativity breaks down before the singularity is reached, but long before that happens the gravitational attraction of the collapsing object becomes so powerful that nothing can escape, not even light. This is where black holes get their name. One way of thinking about what is going on is that the escape velocity from a black hole exceeds the speed of light. Since nothing can travel faster than light, nothing can escape from a black hole.

In fact, any object will become a black hole if it is sufficiently compressed. For any mass, there is a critical radius, called the Schwarzschild radius, for which this occurs. For the Sun, the Schwarzschild radius is just under 3 km; for the Earth, it is just under 1 cm. In either case, if the entire mass of the object were squeezed within the appropriate Schwarzschild radius it would become a black hole.

But although black holes themselves are invisible, they exert a gravitational influence on their surroundings, and this can lead to violent and easily detectable activity in their vicinity. We know that stellar mass black holes exist because some of them are in orbit around ordinary stars, forming binary systems. The direct effect of the gravity of the black hole on the binary orbit of its partner reveals the mass of the black hole, and matter pulled off from the companion streams down towards the black hole, funnelling into its 'throat'. There, the infalling matter gets hot enough to emit X-rays as the particles in the stream speed up and collide with one another.

All these black holes are associated with matter squeezed to very high densities. The black hole at the centre of the Galaxy is a different kind of beast. Curiously, though, such supermassive black holes were the first to pique the curiosity of theorists, long before Albert Einstein came up with the general theory of relativity. In 1783, John Michell, a Fellow of the Royal Society, pointed out that, according to Newton's theory of gravity, an object with a diameter 500 times that of the Sun (about as big across as the Solar System) but with the same density as the Sun would have an escape velocity greater than that of light. (Michell did not use the term 'escape velocity' but in modern language that is what he was talking about; Einstein's theory, of course, makes the same prediction.) This need not involve superdensities at all, since the overall density of the Sun is only about one-and-a-half times the density of water. The Frenchman Pierre Laplace reached the same conclusion independently in 1796, and commented that, although these dark objects could never be seen directly, 'if any other luminiferous bodies should happen to revolve about them we might still perhaps from the motions of these revolving bodies infer the existence of the central ones'. Two centuries later, that is exactly how the black hole at the heart of the Milky Way was discovered.

The centre of the Milky Way lies in the same direction on the sky as the constellation Sagittarius, but much farther away. The constellations, named in ancient times, are patterns of nearby stars, which look bright simply because they are close to us. The names are still used by astronomers to indicate which part of the sky – which direction – an object lies in. That is why M31 is also known as the Andromeda Nebula (or Andromeda Galaxy), even though it is a couple of million light years farther away than the stars in the constellation Andromeda, and has nothing at all to do with them. In the same way, a powerful source of radio noise at the centre of our Galaxy is known as Sagittarius A, even though it has nothing to do with the stars of the constellation Sagittarius.

It only became possible to study the centre of our Galaxy when radio telescopes and other instruments that do not rely on visible light became available. There is a great deal of dust in the plane of the Milky Way, responsible for the interstellar extinction that plagued early attempts at determining the distance scale and providing some of the raw material for new generations of stars. This blocks out visible light. But longer wavelengths penetrate the dust more easily. That is why sunsets are red – short wavelength (blue) light is scattered out of the line of sight by dust in the atmosphere, while the longer wavelength red light gets through to your eyes. So our understanding of the galactic centre is largely based on infrared and radio observations.

More detailed studies showed that Sagittarius A is actually made up of three components lying close to one another. One is the expanding bubble of gas associated with a supernova remnant, one is a region of hot, ionized hydrogen gas, and the third, dubbed Sgr A*, is at the very centre of the Galaxy.

There is certainly plenty of activity around Sgr A*. Infrared studies reveal a dense cluster of stars in which 20 million stars like the Sun are packed into a volume one parsec across, where the stars are on average only a thousand times farther apart than the distance from the Earth to the Sun, and collisions occur every million years or so. There is a massive ring of gas and dust around this cluster, extending out from about 1.5 pc to a distance of 8 pc (some 25 light years), with traces of shock waves from recent explosive events, and both X-rays and even more energetic gamma rays pour out of the central region.

But for all of this high-tech stuff, the best evidence for the presence of the black hole comes from the kind of study Laplace envisaged. Observations made at infrared wavelengths using a telescope with a 10-metre diameter mirror at the Mauna Kea Observatory in Hawaii provided measurements of the speed with which 20 stars close to the galactic centre are moving. The stars are orbiting the galactic centre at speeds of up to 9,000 km per second, which converts to nearly 30 million miles per hour. They are moving so fast that even though they are so far away – nearly 10 kpc – their positions are seen to change in photographs taken at intervals a few months apart over a few years, and by putting such pictures together it is possible to make a movie which actually shows the orbits of the innermost of these stars. The orbital motion tells us that the stars are in the grip of an object with between two and three million times the mass of our Sun. Since this is contained in a volume of space no bigger across than the radius of the Earth's orbit around the Sun, it is definitely a supermassive black hole.

The black hole is relatively quiet today, because it has swallowed up all of the matter in its immediate locality. The activity we can detect now results from a dribble of matter falling into the hole from the surrounding ring of stuff; all it needs to 'eat' each year in order to maintain the present level of activity is a mass equivalent to about 1 per cent of the mass of our Sun, releasing gravitational energy as the matter falls into the hole. Things must have been different long ago when the Galaxy was young and the region around the black hole had not been swept clear of gas and dust; I shall discuss this later, but it is clear that supermassive black holes are the seeds from which galaxies grew.

The way stars move farther out from the galactic centre can also tell us something about the way the Galaxy got to be the way it is today. The orderly structure described so far, with bulge, disc, and halo components, is not quite the whole story. When astronomers look in detail at the compositions of individual stars and the way they are moving, they find that against the background of the many stars that move together in the Milky Way they can pick out long, thin streams of stars which have similar makeup to one another, different from that of the background stars, and are moving in the same direction as each other, at an angle to the motion of most of the stars in that part of the sky.

Nine or ten such streams have now been identified (the exact number depends on how reliable you think the evidence is), with more still being found. They range in mass from a few thousand to a hundred million solar masses of material and in length from 20,000 to a million light years. Very often, these star systems can be traced as tenuous connections to a globular cluster, or to one of the 20 or so small galaxies that orbit the Milky Way Galaxy like moons orbiting around a planet. The most spectacular of these star trails, from our perspective, is called the Sagittarius stream. It extends over a curving span of more than a million light years, and joins the Milky Way to the so-called Sagittarius dwarf elliptical. Another stream, seen in the direction of the constellation Virgo and therefore called the Virgo stellar stream, is moving almost perpendicular to the plane of the Milky Way, and is associated with another dwarf galaxy.

This kind of evidence explains the origin of the star streams. Small galaxies that come too close to our own Galaxy get broken up and dissipated by the gravitational forces – tides – they encounter, trailing a stream of stars as they move in their orbits around the Milky Way. The Sagittarius dwarf is in the final stages of this process, barely discernible today as a coherent group of stars. Eventually, there will be nothing left but the star stream, which will merge with the Milky Way and finally lose its identity.

This is a clear indication that the Milky Way reached its present size through a kind of intergalactic cannibalism, swallowing up its lesser neighbours. Using powerful statistical techniques, astronomers are even able to work backwards from the observations of how star streams are moving today to reconstruct the ghosts of former satellite galaxies, like palaeontologists reconstructing the appearance of a dinosaur from a few fossil remains. And as the icing on the cake, the shape of the orbits of these star streams tells us that the extended halo of dark matter in which the Milky Way is embedded is spherical, not ellipsoidal.

These galactic interactions are not, though, confined to occasions when a large galaxy swallows up its small neighbours. As Vesto Slipher discovered, light from the Andromeda galaxy shows a blueshift corresponding to an approaching velocity of more than 100 km per second (approaching 250,000 miles per hour). The reason it does not show a redshift is because, as Hubble realized, the cosmological redshift is not caused by motion through space. At the distance to the Andromeda galaxy, the cosmological redshift would be tiny, equivalent in velocity terms to less than half the Andromeda galaxy's observed blueshift. But galaxies do move through space, and these motions cause Doppler effects which are superimposed on their cosmological redshifts.

For all but the nearest of our neighbours, the cosmological redshift is much bigger than any Doppler effect, and dominates. But in the case of the Andromeda galaxy, the Doppler effect is much bigger than the cosmological redshift. The Andromeda galaxy really is moving rapidly towards us, and will collide with the Milky Way in about four billion years from now – coincidentally, just when the Sun is nearing the end of its life. Such a collision between roughly comparable disc galaxies will lead to a merger. The stars in each galaxy are separated by such great distances that there will be no collisions between stars in the two discs, but computer simulations show that the gravitational forces will cause the structure of the two discs to be destroyed as the stars merge into one system, forming a giant elliptical galaxy.

All of the discoveries described in this chapter would be important if they only told us about the Milky Way, our island home. But they are doubly important because there is strong evidence that the Milky Way Galaxy is a completely ordinary disc galaxy, a typical representative of its class. Since that is the case, it means that we can confidently use our inside knowledge of the structure and evolution of our own Galaxy, based on close-up observations, to help our understanding of the origin and nature of disc galaxies in general. We do not occupy a special place in the Universe; but this was only finally established at the end of the 20th century.

Chapter 4

Interlude: galactic mediocrity

It can be argued that the scientific revolution began in 1643, with the publication by Nicolaus Copernicus of a book, *De Revolutionibus Orbium Coelestium*, setting out the evidence that the Earth is not at the centre of the Universe, but moves around the Sun. Since then, it has become appreciated that the Sun is just an ordinary star, occupying no special place in our Milky Way Galaxy, let alone in the Universe, and that humankind is just one species of life on Earth, occupying no special place except from our own parochial point of view. With tongue only slightly in cheek, some astronomers say that all of this is evidence in support of the 'principle of terrestrial mediocrity', which says that our surroundings are completely lacking in any special features, as far as the Universe is concerned. This might be a humbling thought for anyone still harbouring pre-Copernican ideas; but if it is correct, it does mean that we can extrapolate from our observations of our surroundings to draw meaningful conclusions about the nature of the Universe at large. If the Milky Way is mediocre, then billions of other galaxies must be very much like the Milky Way, just as one city suburb looks much the same as another city suburb.

But in the decades following Hubble's first measurements of the cosmological distance scale, the Milky Way still seemed like a special place. Hubble's calculation of the distance scale implied that other galaxies are relatively close to our Galaxy, and so they would not have to be very big to appear as large as they do on the sky; the Milky Way seemed to be by far the largest galaxy in the Universe. We now know that Hubble was wrong. Because of the difficulties he struggled with, including extinction and a serious confusion between Cepheids and other kinds of variable stars, the value he initially found for the Hubble Constant was about seven times bigger than the value accepted today. In other words, all the extragalactic distances Hubble inferred were seven times too small. But this was not realized overnight. The cosmological distance scale was only revised slowly, over many decades, as observations improved and one error after another was corrected. I do not intend to take you through all the steps here, but to present the simplest and most direct evidence, using the latest and best observations, for the galactic mediocrity of the Milky Way.

Even in the 1930s, some scientists were unhappy about the idea that the Milky Way might be an unusually large galaxy. The person who felt most strongly about this, and expressed his doubts most forcefully, was the astronomer Arthur Eddington, best remembered as the leader of the eclipse expedition of 1919 that verified the predictions of Einstein's general theory of relativity. Eddington was a firm believer in what is now know as the principle of terrestrial mediocrity, and in his book *The Expanding Universe*, published in 1933, he wrote:

> The lesson of humility has so often been brought home to us in astronomy that we almost automatically adopt the view that our own galaxy is not specially distinguished – not more important in the scheme of nature than the millions of other island galaxies. But astronomical observation scarcely seems to bear this out. According to the present measurements the spiral nebulae, though bearing a general resemblance to our Milky Way system, are distinctly smaller. It has been said that if the spiral nebulae are islands, our own galaxy is a continent. I suppose that my humility has become a middle-class pride, for I rather dislike the imputation

that we belong to the aristocracy of the universe. The Earth is a middle-class planet, not a giant like Jupiter, nor yet one of the smaller vermin like the minor planets. The sun is a middling sort of star, not a giant like Capella but well above the lowest classes. So it seems wrong that we should happen to belong to an altogether exceptional galaxy. Frankly I do not believe it; it would be too much of a coincidence. I think that this relation of the Milky Way to other galaxies is a subject on which more light will be thrown by further observational research, and that ultimately we shall find that there are many galaxies of a size equal to and surpassing our own.

Eddington's argument made complete sense, and it eventually turned out that he was right. But in 1933, this was based on no more than his 'middle-class pride'. After all, some galaxies *are* bigger than others, and if the universe really was dominated by one huge galaxy surrounded by a host of smaller ones you could argue that it was more likely than not that we should find ourselves on the continent rather than on one of the islands. The only way to settle the issue would be to have accurate distance measurements to a large enough number of other disc galaxies to have a good understanding of their sizes in relation to that of the Milky Way. That meant Cepheid distances, and enough of these simply were not available before the launch of the Hubble Space Telescope in 1990 and its repair in 1993.

The importance of determining the cosmological distance scale accurately, more than half a century after Hubble's pioneering work, was still so great that it was a primary justification for the existence of the Hubble Space Telescope (HST). The expressed aim of the Hubble key project was to use the telescope to obtain data from Cepheids in at least 20 galaxies and use them to pin down the value for the Hubble Constant to an accuracy of plus or minus 10 per cent. By the end of the observing phase of the key project, distances to 24 galaxies had been determined accurately using Cepheids. While the Hubble team moved on to the next phase, using these data to calibrate other indicators

8. The Hubble Space Telescope in orbit

such as supernovae, the basic Cepheid data were made available to other astronomers. Together with Simon Goodwin and Martin Hendry, at the University of Sussex, in 1996 I used these Cepheid distances, the 'further observational research' Eddington had called for, to test his belief that the Milky Way is just an ordinary spiral. (The results were published in 1998.)

Using mostly HST data and some from ground-based telescopes, we found that there were 17 spirals, closely resembling the Milky Way in appearance, which had well-determined distances. The standard way to measure the angular diameter of a galaxy is, in effect, to draw contour lines of brightness (isophotes) around it, and to make the cut-off at a certain brightness level. With angular diameters determined in this way and accurate distances from the Cepheids, the true linear sizes of the 17 galaxies followed.

The hardest part of the project turned out to be measuring an equivalent diameter for the Milky Way – the classic problem of the difficulty of seeing the wood for the trees. But observations of the distribution of stars within the Milky Way made it possible to work out what it would look like from above, and this gave us an equivalent isophotal diameter of just under 27 kiloparsecs. The big question was, how would this compare with the diameters of the other 17 galaxies? The short answer is that the average diameter of all the 18 galaxies in our sample, including the Milky Way, was just over 28 kpc. Exactly as Eddington had surmised, the Milky Way is an ordinary spiral, with a diameter fractionally, but not significantly, smaller than the average. Most definitely, it is *not* a continent among islands. But nor is it significantly smaller than average. The Milky Way is, in a word, mediocre.

Among other things, this makes it possible to use observations of galaxy diameters to determine the value of the Hubble Constant, and to do so within the 10 per cent accuracy set as a target by the Hubble key project. By putting this in a cosmological context, as I shall do in the next chapter, it reveals the age of the Universe itself – the time that has elapsed since the Big Bang.

Chapter 5
The expanding universe

Modern cosmology began with Hubble's two great discoveries about galaxies – that they are other islands in space outside the Milky Way, and that there is a relationship between the redshift in the light from a distant galaxy and its distance. Together, these two discoveries mean that galaxies can be used as test particles to reveal the overall behaviour of the Universe. In particular, they show that the Universe is expanding.

Although the discovery of the redshift–distance relationship came as a surprise at the end of the 1920s, it was almost immediately realized that a mathematical theory describing this kind of universal behaviour had already been found – Albert Einstein's general theory of relativity. The general theory describes the relationships between space, time, matter and gravity. One of the key features of the theory is that space and time should not be thought of as separate entities, but as facets of a single four-dimensional entity known as spacetime. The idea of four-dimensional spacetime dates back to 1908, when Hermann Minkowski refined Einstein's special theory of relativity, which he had published in 1905. 'Henceforth,' Minkowski said, 'space by itself, and time by itself, are doomed to fade into mere shadows, and only a kind of union of the two will preserve an independent reality.'

The limitation of the special theory (the reason why it is 'special', as in a special case of something more general) is that it does not deal with gravity or with acceleration. It describes precisely the relationships between all moving objects and light (used as a general term for all electromagnetic radiation) as long as they are moving in straight lines at constant speed, and how the world would look from the point of view of any of those objects. These was a far greater achievement than such a quick summary suggests, because Einstein had in effect modified Isaac Newton's understanding of dynamics to take account of James Clerk Maxwell's understanding of light. But it was only intended as an interim step on the road to a complete theory which included gravity and acceleration as well.

Einstein achieved that with the general theory, which he completed in 1915. The simplest way to understand the general theory is in terms of Minkowski's four-dimensional spacetime. Einstein discovered that spacetime is elastic, so it is distorted by the presence of matter. Objects moving through spacetime follow curved paths around the distortions caused by the presence of matter, rather like the way a marble rolled across a trampoline will follow a curved path around the indentation made by a heavy object, such as a bowling ball, placed on the trampoline. The effect we call gravity is a consequence of the curvature of spacetime. In a famous aphorism, 'matter tells spacetime how to bend, spacetime tells matter how to move'.

Crucially, light rays also follow the appropriate curved paths through spacetime in the presence of matter. The effect is very small, unless the amount of matter involved is large, or it is squeezed into a small volume at very high density, or both. But it is just detectable in the region of space near the Sun. The general theory predicted that light from distant stars passing close to the edge of the Sun would be bent by a certain amount because of the way the Sun's mass distorts spacetime in its vicinity. From Earth, the effect would be to shift the apparent positions of the

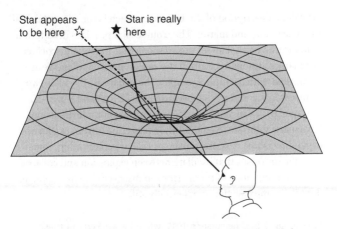

Star appears to be here ☆   Star is really here ★

9. The Sun distorts spacetime in its vicinity, like the dent made by heavy object placed on a trampoline. Light from a distant star follows the curve in space, so the star appears to be shifted from its position when the Sun is not in the line of sight

background stars, compared with observations of the same part of the sky made when the Sun was not in the way. Since the background stars cannot be seen against the glare of the Sun, the only way to observe these changes would be during a total solar eclipse, when the Sun's light is blocked by the Moon. By great good fortune for astronomers, a suitable eclipse occurred in 1919. This was the occasion when a team led by Arthur Eddington measured the effect and found that it exactly matched the predictions of Einstein's theory; it was from that moment that Einstein became a famous celebrity, although many people never quite knew what he was famous for. Since then, the general theory has passed every test that has been devised, most recently a subtle experiment flown into space to monitor the effects of the Earth's gravity on weightless gyroscopes.

The general theory of relativity is the best theory we have to describe the overall behaviour of space, time and matter. As Einstein realized from the outset, this means that it automatically provides a description of the Universe, which is the sum total of all the space, time and matter. The trouble is, it provides descriptions of many universes. The set of equations that Einstein discovered has many solutions, as is often the case in mathematics. We are all familiar with a simple example. The equation $x^2 = 4$ has two solutions, $x = 2$ and $x = -2$, because both (2×2) and (-2×-2) are equal to 4. Einstein's equations are more complicated, and have many solutions. Some solutions describe universes that are expanding, some describe universes that are contracting, some describe universes that oscillate between expansion and collapse, and so on. But none of them, Einstein discovered to his surprise, describe a universe that is essentially still.

He was surprised because in 1917, when he worked out these solutions after completing the general theory, everyone thought that the Universe was static. The Milky Way was still thought by most astronomers to be essentially the entire universe, and although stars move around within the Milky Way, overall it is neither expanding nor contracting. The only way Einstein could obtain a mathematical description of a static universe within the framework of the general theory of relativity was to introduce an extra term into the equations, now known as the cosmological constant and usually represented by the Greek letter lambda (Λ). A dozen years later, when Hubble discovered the redshift–distance relationship, it turned out to match the mathematical description of an expanding universe in one of the simplest solutions to Einstein's equations, without the lambda term. Einstein described the introduction of the cosmological constant as 'the biggest blunder' of his career, and it was discarded by everyone except a few mathematicians who liked playing with equations for their own sake, whether or not they describe the real Universe.

The full implications of the discovery that the general theory of relativity provides a good description of our Universe are explained in Peter Coles's book. The key point to grasp, though, is that the expansion described by the equations is an expansion of

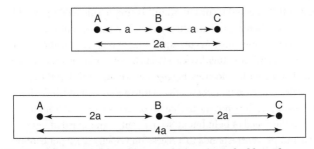

10. Expanding spacetime is like stretching a piece of rubber. The 'galaxies' A, B, and C do not move through the space between them. But when the space expands to double the distance between A and B, it also doubles the distance between every other pair of galaxies, including A and C. From the viewpoint of every galaxy in this universe, every other galaxy is receding at a rate which is proportional to its distance. Because C is twice as far away from A as B is, for example, when all distances are doubled (when the scale factor doubles) it seems that C has 'moved away' from A twice as fast as B has

space as time passes. The cosmological redshift is not a Doppler effect caused by galaxies moving outward through space, as if fleeing from the site of some great explosion, but occurs because the space between the galaxies is stretching. So the spaces between galaxies increase while light is on its way from one galaxy to another. This stretches the light waves to longer wavelengths, which means shifting them towards the red end of the spectrum.

The way the stretching occurs, though, produces redshifts which depend on relativistic effects. If we translate the redshifts into equivalent velocities, then provided the velocities involved are small compared with the speed of light, they behave in a very simple way. Redshifts are usually denoted by the letter z. If $z = 0.1$, that means an object is receding at one tenth of the speed of light (i.e. about 30,000 km per second, bigger than anything measured in the pioneering study by Hubble and Humason). A redshift of 0.2 means it is receding twice as fast, and so on – up to a point. Since nothing can travel faster than light, the largest redshift that could be produced if this simple rule held up would be 1. But when relativistic effects are taken into account, the largest possible redshift, corresponding to recession at the speed of light, is infinite. Relativistic effects become important once we are dealing with 'velocities' bigger than about a third of the speed of light. Once we take these effects into account, a redshift of 2, for example, does not mean that an object is receding from us at twice the speed of light but at 80 per cent of the speed of light; a redshift of 4 corresponds to a recession velocity of 92 per cent of the speed of light. Individual redshifts greater than 10 have now been measured, but these are very much the exception.

In fact, there are very few isolated galaxies in the Universe. Most galaxies occur in clusters, which may contain anything from a few galaxies to thousands of galaxies, held together by gravity. Individual galaxies within the cluster are moving around their mutual centre of mass, while the whole cluster is being carried along by the expansion of space. Like a swarm of bees, the galaxies move around one another while the whole swarm moves along as a unit. So when we look at the light from galaxies in a cluster, we find that there is some average redshift, which is the cosmological redshift caused by the expansion of the Universe, but that some galaxies have slightly bigger redshifts and some slightly smaller redshifts. The galaxies with smaller redshifts are the ones that are moving towards us, so that their motion *through* space contributes a Doppler blueshift which reduces the overall redshift. The galaxies with larger redshifts are the ones that are moving away from us, so that their motion *through* space contributes a Doppler redshift which enhances the overall redshift. All of is this taken account of when astronomers use the shorthand expression 'galaxies show a redshift proportional to their distance'.

The second key point about the universal expansion is that it does not have a centre. There is nothing special about the fact that we observe galaxies receding with redshifts proportional to their distances from the Milky Way. In another example of terrestrial mediocrity, whichever galaxy you happen to be sitting in, you will see the same thing – redshift proportional to distance. A simple analogy makes this clear. Imagine the surface of a perfect sphere, painted with random spots of colour to represent galaxies. If you inflate the sphere, the distances between every spot of paint increase, in exactly the same way that separations between galaxies increase in the real Universe as it expands. Suppose that the expansion doubles the distance between each spot of paint. Spots that were two centimetres apart end up four centimetres apart; spots that were four centimetres apart end up eight centimetres apart, and so on. If before the expansion there were three spots spaced two centimetres apart in a straight line, then after the expansion the distance from the central spot to either of its neighbours will now be four centimetres, but the distance between the two outer spots will now be eight centimetres. From either of the end spots, the central spot will have receded by two centimetres, but the other end spot will be seen to have receded by four centimetres. It started out twice as far away as the central spot, and the amount of its 'redshift' is twice as big as for the nearer spot. From every spot on the surface of the sphere, the overall picture is the same. Redshift is proportional to distance.

But what if we imagine reducing the size of the sphere? Now, the spots get closer together, and 'blueshift' is proportional to distance. This is equivalent to looking back in time to the history of the expanding Universe. It is obvious that if galaxies are moving apart today then they must have been closer together in the past. It is considerably less obvious, but required by the general theory of relativity, that if you wind this expansion backwards from how things are today for long enough you reach a time when all of the matter and all of space were merged into a mathematical point, a singularity, with zero volume and infinite density, like the singularities predicted to lie at the hearts of black holes. As with black hole singularities, because physicists do not believe theories that predict infinitely extreme physical conditions, it is thought that the general theory must break down when pushed that far.

But there is every reason to believe that the Universe started from a state of extremely small volume (smaller than an atom) and extremely high temperature and density (containing all the mass in the Universe today), even if none of these properties was ever infinite. This idea of a superdense, superhot beginning is the core of the Big Bang model of the Universe. The idea of the Big Bang began to be taken seriously in the second half of the 20th century, as more observations confirmed the reality of the universal expansion. The big question which cosmologists struggled to answer was, when did the Big Bang happen? How old is the Universe? The answer came from studies of galaxies, providing measurements of the Hubble Constant.

The Hubble Constant is a measure of how fast the Universe is expanding today. If it has always been expanding at the same rate, that tells us how long it has been since the Big Bang. Take 1 divided by the value of the Hubble Constant $(1/H)$ and you know how long it is since the galaxies were on top of each other – the time since the Big Bang. In the same way, if a car leaves London heading west along the M4 at a steady 60 miles an hour, when it is 120 miles from London we know that the journey started two hours ago. Things are slightly more complicated because the simplest model of the Universe derived from Einstein's equations says that it must have started out expanding more rapidly, and slowed down as time passed, because of gravity holding back the expansion. A better estimate for the age of the Universe is two-thirds of $(1/H)$, and $1/H$ itself is referred to now as the Hubble time. But the crucial point is that if we can measure H we can measure the age of the Universe.

Because the age is inversely proportional to H, the smaller the value of the Hubble Constant the older the Universe must be. Using Hubble's own value for the constant, 525 kilometres per second per Megaparsec, the age of the Universe comes out as about two billion years. Even in the 1930s, it was clear that something was wrong with this estimate, because it is less than the age of the Earth. This is one reason why the idea of the Big Bang only began to be taken seriously after the 1940s, when there was a drastic revision of the distance scale after the confusion between different kinds of variables was ironed out. At a stroke, the Hubble Constant was halved and the estimated age of the Universe doubled, making the Universe seem to be about as old as the Earth.

But at about the same time, astronomers began to develop a good understanding of how stars work, and to derive reliable estimates of their ages. Some stars turn out to be more than ten billion years old, which again provided embarrassment for the Big Bang idea as it stood in the 1950s. This was one reason why a rival cosmology, the Steady State model, seemed attractive to some astronomers at the time. The idea behind the Steady State model was that as the galaxies moved apart in the expanding universe, the forces responsible for the stretching of space also caused the appearance of new matter in the gaps between the galaxies – atoms of hydrogen that would form clouds of gas from which new galaxies would form to fill the gaps. On that picture, there had been no beginning and there would be no end, with the universe always having much the same appearance overall. The death knell of the Steady State model was sounded in the 1960s, when radio astronomers discovered a weak hiss of radio noise coming from all directions in space. This cosmic microwave background radiation, which had been predicted by Big Bang theory (although the prediction had been forgotten!), is interpreted as the fading remnant of the energetic radiation from the Big Bang itself, an interpretation reinforced by later observations, including those from dedicated satellites sent in to space to study it. The need for the Steady State alternative also declined because estimates for the age of the Universe gradually increased as the years passed.

From about 1950 onwards, gradual revisions of the distance scale based on ever-improving observations pushed the value of the Hubble Constant down until, by the beginning of the 1990s, it was known to lie somewhere in the range from 50 to 100, in the usual units. As an astronomer would put it, 75 ± 25. This was where the Hubble key project came in.

Like the Andromeda galaxy, galaxies in clusters typically have random motion through space of a few hundred kilometres per hour. This means that in order to get reliable estimates of the cosmological redshift of a cluster it is best to look at distant clusters, where the cosmological redshift is greater and individual random velocities and their associated Doppler shifts are a smaller proportion of the overall redshift. But, of course, it is harder to measure distances for more distant clusters, so there is a trade-off when it comes to using clusters in this way to determine the value of the Hubble Constant. The Hubble key project used the traditional technique devised by Hubble himself of getting accurate distances to nearby galaxies from Cepheids, using these Cepheid distances to calibrate the brightness of other indicators such as supernovae, and moving out into the Universe in a series of steps. The difference was, working 60 years after Hubble, they had a better telescope, the confusion between different kinds of variable stars had been resolved, extinction was understood, and the secondary indicators such as supernovae were also much better understood than in Hubble's day. The final estimate that the key project team came up with for H, in May 2001, was 72 ± 8, corresponding to an age of the Universe of about 14 billion years. Happily, during the previous decade, the 1990s, the ages of the oldest stars we can see were determined by quite independent techniques to be around 13 billion years. The Universe really is older than the stars and galaxies it contains.

This is a much more profound result than it might seem at first sight. The age of the Universe is determined by studying some of the largest things in the Universe, clusters of galaxies, and analysing their behaviour using the general theory of relativity. Our understanding of how stars work, from which we calculate their ages, comes from studying some of the smallest things in the

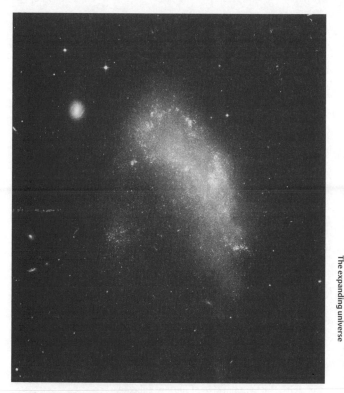

11. The irregular galaxy NGC 1427

Universe, the nuclei of atoms, and using the other great theory of 20th-century physics, quantum mechanics, to calculate how nuclei fuse with one another to release the energy that keeps stars shining. The fact that the two ages agree with one another, and that the ages of the oldest stars are just a little bit less than the age of the Universe, is one of the most compelling reasons to think that the whole of 20th-century physics works and provides a good description of the world around us, from the very small scale to the very large scale.

A value for the Hubble Constant close to 70 kilometres per second per Megaparsec has now been confirmed by other independent techniques. Some of these involve high-tech equipment such as satellites and a sophisticated understanding of physics; but one simple approach highlights the relationship between galaxies and the Universe, and, when combined with the more sophisticated measurements, provides a confirmation of our mediocrity.

The evidence that the Milky Way is just an average spiral is based on a fairly small sample of galaxies fairly close to us, in cosmological terms. If we accept this at face value, though, it provides us with a way to estimate the distances to other galaxies, by comparing their sizes with the size of the Milky Way, or with the average of our local sample, which is near enough the same thing. There is little point in making such comparisons with individual galaxies, because we know that there is a wide range of sizes. The largest spiral galaxy in our cosmic neighbourhood, M101, has a diameter of nearly 62 kpc, more than twice that of the Milky Way, so estimating its distance by assuming it is the same size as our Galaxy would not be a good idea. What we need is some kind of statistical measurement so that we can take the average size of galaxies far away across the Universe and compare that with the average size of nearby galaxies.

Since Hubble's day, observers have built up catalogues giving the positions, redshifts, and angular sizes of thousands of galaxies – many different catalogues each containing thousands of galaxies. Some of these include angular sizes, often given in terms of the same isophotal diameters used to determine the mediocrity of the Milky Way. Each angular diameter can be converted into a true linear diameter by multiplying it by a number which depends only on the redshift, which we know, and H, which we are assuming we do not know. If we take thousands of galaxies with different redshifts, scattered all over the sky, it is possible to choose some value of H and work out all of the linear diameters,

12. The central region of the galaxy M100, imaged by WFPC2 on the HST

then take an average over the whole sample to estimate the average size of a galaxy. It is straightforward to do this over and over again using a computer which keeps on varying H until the average value that comes out of the calculation is the same as the average diameter of the nearby spirals like the Milky Way. This gives a unique value for H.

There are practical difficulties that have to be overcome. Among other things, you have to make sure that all of the diameters have been measured in the same way, that the sample is restricted to galaxies which have the same overall structure as the galaxies in our local sample, and that the observations are indeed picking up all of the relevant galaxies. One of the most important factors to allow for is that it is easier to see bigger galaxies, so for larger redshifts there will be fewer small galaxies than there should be in the sample because they have been overlooked. This is an effect known as Malmquist bias. Fortunately, by comparing the numbers of galaxies of different sizes at different redshifts it is possible to work out the statistics of this effect – the way small galaxies drop out of the sample as redshift increases – and correct for it. In a further complication, nearby galaxies have to be left out of the calculation, because their random Doppler shifts are comparable to their cosmological redshifts and confuse the picture. But the technique works for galaxies out to about 100 Megaparsecs, and even with all these restrictions one of the standard catalogues, known as RC3, provides a sub-set of well over a thousand suitable galaxies that satisfy all these criteria. This is ample to provide a statistically reliable sample. When all of the work is done, the value for the Hubble Constant based on comparison of galaxy diameters comes out in the high 60s, *if* the Milky Way is indeed just an average spiral. This value agrees with the other measurements.

This is far from being the best or most accurate way to measure the Hubble Constant, but it is valuable for two reasons. First, it is a nice, physical technique which can be understood in terms of our everyday experience, where we know that a cow standing on the other side of a large field only looks small because it is so far away. It does not require any deep understanding of physics or mathematics. Second, the argument can be turned on its head. The first real proof that the Milky Way is just an average spiral came from comparing its size with the sizes of just 17 relatively nearby galaxies. But if *H* is close to 70, as the more sophisticated observations and analyses indicate, then we can use that value to calculate the average size of the 1,000-plus galaxies in our sample, some of them 100 Mpc away, and we find that it is indeed very close to the size of the Milky Way and the average size of our nearby sample. At the very least, our Galaxy is typical of the kind of disc galaxy found in a 'local' region of space some 200 Mpc across with a volume of more than 4 million cubic Megaparsecs.

But this is indeed still a local bubble compared with the size of the observable Universe. There are objects known with measured redshifts corresponding to distances greater than ten billion light years, 30 times farther away than the most distant galaxies used in this technique for estimating the value of H. Studies of these objects show that there is more to the story. It seems that the expansion of the Universe has not slowed down since the Big Bang in the way predicted by the simplest solutions to Einstein's equations, but that it may have begun to speed up.

In the 1990s, astronomers began to use supernova observations to calibrate the redshift–distance relationship for redshifts of about 1 (the largest known redshifts for such supernovae are less than 2). The technique depends on the discovery that a certain kind of supernova, a family known as SN1a, all seem to peak at the same absolute brightness. This was discovered from observations of SN1a in nearby galaxies for which distances are now very well known. The discovery was particularly important because supernovae are so bright that they can be seen at very great distances.

Although all SN1a have the same absolute brightness, the farther away they are across the Universe the fainter they look. This means that if they really do all reach the same absolute peak brightness, by measuring the apparent peak brightness of SN1a in very distant galaxies we can work out how far away those galaxies are. If we can also measure redshifts for the same galaxies, we can calibrate the Hubble Constant. When these observations, at the very limit of what was technologically possible, were carried out, the observers found that the supernovae in very distant galaxies are a little fainter than they should be, if the galaxies in which they

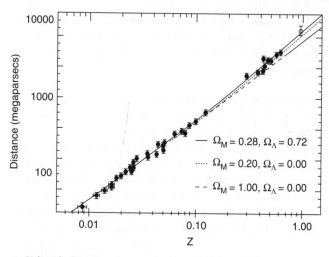

13. **Using observation of supernovae at very high redshifts, the redshift–distance plot can be extended far out into the Universe. The best fit to the data (solid line) includes an allowance, Λ, for the cosmological constant described previously in the text**

reside are at the distances indicated by the accepted value of the Hubble Constant.

The possibility that supernovae in such distant galaxies really do not shine as brightly as those in galaxies closer to us cannot be ruled out, but the conclusion that best fits all of the available evidence is that these supernovae are a little bit farther away from us than they should be if the Universe has been expanding in line with the simplest cosmological models ever since the Big Bang. Just one tiny modification to Einstein's equations is needed to make everything fit – a small cosmological constant has to be put back in to the equations. Perhaps it wasn't such a blunder after all.

When Einstein introduced his cosmological constant, he did so to hold the model universe still. But different choices of the constant will make it expand faster or slower, or even make it collapse. The presence in the equations of the kind of cosmological constant required to explain the supernova observations implies that the entire Universe is filled with a kind of energy which has no noticeable local effect on everyday matter, but acts like a kind of compressed elastic fluid, pushing the Universe outward and acting against the inward pull of gravity. Because the cosmological constant is traditionally labelled Λ, this is known as the lambda field. With a suitable choice of density for this field, it is straightforward to explain how the expansion of the Universe slowed down for the first few billion years after the Big Bang, as the simpler models predicted, but then began very slowly to speed up.

It works like this (there are more complicated possible explanations for the cosmic acceleration, but since the simplest explanation works beautifully I shall not discuss them here). The lambda field is constant, and has had the same value since the Big Bang. Because we cannot see this field, it is often called 'dark energy'. Dark energy is a property of spacetime itself, so when space stretches and there are more cubic centimetres to fill, the dark energy does not get diluted. This means that the amount of energy stored in every cubic centimetre of space stays the same, and it always exerts the same outward push in every cubic centimetre. This is quite different to what happens to matter as the Universe expands. As the Universe emerged from the Big Bang, the density of matter was as great everywhere as the density of an atomic nucleus today. A thimbleful of such material would contain as much mass as all the people on Earth today put together, and so the gravity associated with that density of matter completely overwhelmed the lambda field. As time passed, the Universe expanded and the same amount of matter occupied an increasing volume of space; the density of matter declined accordingly. This meant that the gravitational influence on the expansion gradually got smaller, until it was less than the influence of the dark energy.

To explain the supernova observations, the influence of matter on the expansion, acting to slow the expansion down, must have weakened to the point where it was the same size as the influence of dark energy, acting to make the expansion speed up, about five or six billion years ago. In redshift terms, the switch took place between a redshift of 0.1 and a redshift of 1.7. Since then, the influence of dark energy has been bigger than the influence of matter, making the expansion of the Universe accelerate.

If the expansion is accelerating, one implication is that the Universe is very slightly older than the 14 billion years calculated assuming no acceleration, because if the Universe was expanding more slowly in the past it would have taken longer to reach its present state. But this effect is very small, and it works in the right direction to keep the age of the Universe bigger than the ages of the oldest stars, so it need not concern us here.

The amount of dark energy required to do all this is tiny. Bearing in mind Einstein's discovery that energy and mass are equivalent to one another, the amount of mass associated with dark energy is a bit less than 10^{-29} grams in every cubic centimetre of the Universe – that is, 0.00000000000000000000000000001 grams in every cubic centimetre. So it cannot make the Earth, or the Solar System, or the Milky Way, or even a cluster of galaxies expand and break apart, because on a local scale the gravity of concentrations of matter completely overwhelms it.

On a cosmic scale, though, the presence of even this much energy, and its equivalent mass, in *every* cubic centimetre of the Universe, even in all the 'empty space' between the stars and galaxies, adds up dramatically. It means that there is far more mass in the form of dark energy than there is in the form of bright stars and galaxies. This would have come as a big surprise to Hubble and his contemporaries, who imagined that they were studying the most important components of the Universe. But at the end of the

1990s it was just what the doctor ordered. By then, it had already become clear that there is more to the Universe than meets the eye, and cosmologists were already trying to find what they called the 'missing mass'. The lambda field turned out to be the missing piece that completed the modern picture of the Universe, which provides a framework within which to understand the origin and evolution of galaxies – which are, after all, still very important for life forms like ourselves.

Chapter 6
The material world

What are galaxies made of? The obvious constituents are hot, bright stars and cool, dark clouds of gas and dust. This is essentially the same kind of material that the Earth is made of, and our own bodies are made of – atomic material. Atoms consist of dense nuclei, composed of protons and neutrons, surrounded by clouds of electrons, with one electron in the cloud for each proton in the nucleus. Inside stars, the electrons are stripped from the nuclei to make a form of matter known as plasma, but it is still essentially the same sort of stuff. Protons and neutrons are members of a family of particles collectively known as baryons, and the term 'baryonic matter' is often used by astronomers to refer to the stuff that stars, gas clouds, planets and people are made of. Electrons are members of a different family, known as leptons. But since the mass of an electron is less than one-thousandth of the mass of either a proton or a neutron, in terms of mass, baryons dominate this kind of familiar matter.

One of the remarkable achievements of modern cosmology is that it is able to tell us how much baryonic matter there is in the Universe – or rather, what the density of such matter, averaged over the entire visible Universe, must be. Drawing on the general theory of relativity, cosmologists measure such densities in terms of a parameter labelled with the Greek letter omega (Ω), which is related to the overall curvature of space. This is most easily understood by making an analogy between the three-dimensional curvature of space and the way a two-dimensional surface can be curved. The surface of the Earth is an example of a closed surface, which is bent around on itself. On such a closed surface, if you travel in the same direction for long enough you get back to where you started. The shape of a saddle is an example of an open surface, which can be extended off to infinity in all directions. Exactly in between these two possibilities there is a flat surface, like the top of my desk, which is not curved at all. Einstein's equations tell us that, depending on how much matter it contains, the shape of our three-dimensional space may either be closed, in the same sense that the two-dimensional surface of a sphere is closed, open, like a saddle surface, or flat, like the top of my desk. A flat universe corresponds to having a value of 1 for the density parameter Ω. A closed universe requires a higher density of matter, an open universe a lower density of matter. Cosmologists measure densities as fractions of this parameter. For example, if the amount of baryonic matter in the Universe were half the

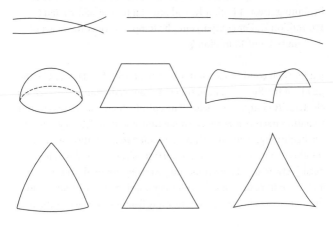

CLOSED   FLAT   OPEN

14. Space may conform to one of three basic geometries. These are represented here by their equivalents in two dimensions

amount required to make the Universe flat (which it is *not*) then we would say Ω(baryon) = 0.5.

All the baryonic matter in the Universe was manufactured in the Big Bang, ultimately out of pure energy in line with $E = mc^2$, which can, of course, be rewritten as $m = E/c^2$. The calculation of the amount of baryonic matter produced in the Big Bang is very straightforward, provided that we can be sure that the temperature of the Big Bang was at least a billion degrees. The evidence for this comes from the weak hiss of radio noise that can be detected coming from all directions in space. This background radio noise is interpreted as the leftover radiation from the Big Bang fireball itself, redshifted by a factor of a thousand so that it now shows up as microwave radiation with a temperature of 2.7 degrees above absolute zero (2.7 K). From the observations, we can work backwards to calculate the temperature of the Universe at any time in the past, when it was smaller and the radiation was correspondingly less redshifted. One second after the beginning of time, the temperature was 10 billion K, 100 seconds after the beginning it was 1 billion K, and after an hour it had cooled to 170 million K. For comparison, the temperature at the heart of the Sun is about 15 million K.

Under such conditions, matter is in the form of a plasma, like the inside of the Sun, and radiation gets bounced around between the electrically charged particles. The cosmic microwave background radiation itself comes to us from a time about 300,000 years after the beginning, when the Universe cooled to a few thousand K, roughly the temperature of the surface of the Sun today. Then, negatively charged electrons and positively charged protons got locked up in neutral atoms and the radiation could stream away through space, just as it streams away from the surface of the Sun.

Conditions in the later stages of this cosmic fireball were very similar to the conditions inside exploding nuclear bombs, which have been studied on Earth. Armed with an understanding of how nuclear explosions work, cosmologists can calculate that the baryonic mix that emerged from the Big Bang was about 75 per cent hydrogen and 25 per cent helium by weight, with just a tiny trace of lithium. But from the way the baryonic particles interact with light under extreme conditions, and measurements of the background radiation, they can also calculate that the total amount of baryonic material produced in the Big Bang and present in the Universe is only 4 per cent of the density required for flatness. In other words, $\Omega(\text{baryon}) = 0.04$.

The obvious next step is to compare this prediction of the amount of baryonic matter in the Universe with the amount we can see in bright stars and galaxies. This is a rough and ready calculation based on our understanding of the brightnesses and masses of stars and the number of stars in galaxies, but it suggests that about a fifth of the baryonic matter, less than 1 per cent of the total amount of matter needed to make the Universe flat, is in the bright stuff. Some of the other four-fifths is in the clouds of gas and dust between the stars, or perhaps in the form of dead, burnt-out stars. Some of it is in the form of a kind of transparent fog of hydrogen and helium surrounding galaxies like our own. And yet, as I mentioned earlier, we know from the way galaxies rotate and the way they move through space that they are held in the grip of a great deal more matter than this. This can only be a form of cold, dark, non-baryonic matter, made up of some kind of particle or particles that have never been detected in any experiment on Earth. It is dubbed Cold Dark Matter, or CDM for short, and detecting it is one of the most pressing tasks of particle physicists today.

Evidence for CDM comes from the way galaxies move – how they rotate, and how they move through space. The rotation of a disc galaxy can be measured using the familiar Doppler effect, which shows how stars on one side of a galaxy are moving towards us as the galaxy rotates, while stars on the other side are moving away from us. This only works for galaxies seen nearly edge-on,

15. The microwave map of the sky obtained by WMAP

but there are plenty of those to study. The Doppler effect adds to the redshift on one side of the disc, and subtracts from it on the other side, so the measured redshift at different places along the disc shows how the stars are moving around the centre of the galaxy. The crucial point is that outside the central nucleus of a disc galaxy, where other interesting things happen, the rotation speed is constant all the way out to the edge of the visible disc. All the stars in the disc are moving at the same speed in terms of kilometres per second. This is quite different from the way the planets of the Solar System move in their orbits around the Sun.

Planets are small objects orbiting a large central mass, and the gravity of the Sun dominates their motion. Because of this, the speed with which a planet moves, in kilometres per second, is inversely proportional to the square of its distance from the centre of the Solar System. Jupiter is farther from the Sun than we are, so it moves more slowly in its orbit than the Earth, as well as having a larger orbit. But all the stars in the disc of a galaxy move at the same speed. Stars farther out from the centre still have bigger orbits, so they still take longer to complete one circuit of the galaxy. But they are all travelling at essentially the same orbital speed through space.

This is exactly the pattern of behaviour that corresponds to the orbital motion of relatively light objects embedded within a large amount of gravitating matter, like raisins moving around inside a loaf of raisin bread. The natural conclusion is that disc galaxies, including the Milky Way, are rotating inside much larger clouds, or haloes, of unseen dark matter. This is some form of spread-out material, so it must be in the form of particles rather like the particles of a gas, which have mass and influence everyday matter gravitationally, but do not interact with everyday matter in any other way (for example, through electromagnetism) or they would have been noticed. On this picture, CDM particles are present everywhere, including the place where you are reading this, and

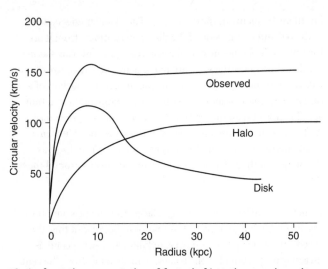

16. A schematic representation of the typical 'rotation curve' seen in a disc galaxy

are continually passing through your body without affecting it. There are thousands, perhaps tens of thousands, of CDM particles in every cubic metre of everything, as well as in every cubic metre of 'nothing' – so-called empty space.

Cold Dark Matter also reveals its presence through its influence on clusters of galaxies. The invaluable Doppler shift can be used to tell us the way an individual galaxy in a cluster is moving relative to the centre of the cluster, and the range of speeds of all the galaxies in a cluster. The clusters can only exist because they are held together by gravity – otherwise, the expansion of the Universe would pull them apart and spread the galaxies out through space. But there are limits to how effective this gravitational constraint can be. If you throw a ball up into the air, it will reach a certain height, depending on its speed, then fall back as the Earth's gravity tugs on it. But if you could throw the ball hard enough it would escape from the Earth entirely and carry on out into space. The minimum vertical speed needed to do this is called the escape velocity, and depends only on the mass of the object you are trying to escape from and how far you are from the centre of the mass. At the surface of the Earth, the escape velocity is 11.2 km per second. If we add up the masses of all the galaxies in a cluster, inferred from their brightnesses *and* including an appropriate allowance for their dark matter haloes, we can work out the escape velocity from the cluster. It turns out that in order for clusters to maintain their gravitational grip on their galaxies there must be even more dark matter in the 'empty space' between the galaxies, as well as the dark matter in the haloes of individual galaxies. The whole Universe is filled with an invisible fog of CDM.

Putting all of the evidence together, it can be calculated that there is nearly six times as much Cold Dark Matter in the Universe as there is baryonic matter. In other words, $\Omega(\text{CDM}) = 0.23$. Adding this to the known amount of baryonic matter in the Universe, we find that 27 per cent of the amount of matter needed to make the Universe flat has been accounted for. That is, $\Omega(\text{matter}) = 0.27$.

This could have been embarrassing for cosmologists, because by the time these calculations were being refined to the accuracy I have given here, around the end of the 20th century, there was other evidence that the Universe is actually flat. It came from studies of the cosmic microwave background radiation, made by instruments carried on balloons and satellites above the obscuring layers of the Earth's atmosphere. Such instruments are now so sensitive that they can pick out variations in the temperature of the radiation from place to place on the sky, looking at hot and cold spots (relatively speaking) that were imprinted on it when the Universe was a few hundred thousand years old.

Before the Universe cooled to the point where electrically neutral atoms could form, radiation and the electrically charged particles of matter were coupled with one another in such a way that differences in the density of matter at different places in the Universe were associated with differences in the temperature of the radiation. About 300,000 years after the Big Bang, when the Universe cooled to the critical temperature, radiation and matter decoupled, and the radiation was left imprinted with a pattern of hot and cold spots corresponding to the pattern of density variations in baryonic matter at that time – a kind of fossil of the large-scale distribution of baryons at the time of decoupling. Because light travels at a finite speed, in 300,000 years it can only travel a distance of 300,000 light years, so in the time from the Big Bang until decoupling the largest regions of the Universe that could have had any kind of internal coherence grew to be 300,000 light years across. This means that the biggest uniform patches seen in the microwave background map of the sky correspond to patches of the Universe that were 300,000 light years across at the time of decoupling.

Since that time, the radiation has streamed across space without interacting directly with matter. But it has been influenced by the curvature of space. We know that a massive object like the Sun bends light passing near its edge. This is very similar to the way a lens bends light. Lenses can make images of distant objects seem bigger (like looking through a telescope) or smaller (like looking through the wrong end of a telescope). So can curved spacetime, depending on the nature of the curvature. Using the general theory of relativity, it is possible to calculate how big the largest uniform blobs in the background radiation should look to our instruments today, if they were 300,000 light years across at the time of decoupling. The observed size depends on the exact curvature, but if the Universe is open we should see a magnification and if it is closed we should see smaller blobs. If it is flat, there should be no effect. The measurements show that the Universe is almost certainly flat, but might *just* be closed. In other words, $\Omega = 1$.

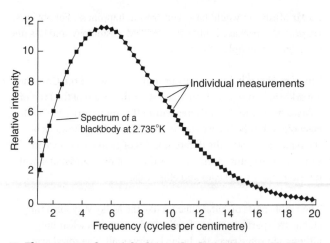

17. The spectrum of cosmic background radiation measured by the COBE satellite

And yet, we know that the total amount of matter in the Universe is less than one-third of the amount needed to make the Universe flat. It could indeed have been embarrassing. But just at the time cosmologists were beginning to worry about this puzzle, the supernova studies came along and showed that the expansion of the Universe is accelerating. The amount by which it is accelerating requires a cosmological constant Λ with a certain strength. This corresponds to a mass density equivalent of 73 per cent of the density of matter needed to make the Universe flat. In other words, $\Omega(\Lambda) = 0.73$. This was just what was needed. Far from being an embarrassment, the discovery that $\Omega(\text{matter}) = 0.27$ turns out to be a triumph. When everything is taken into account, we are left with an equation that is very simple, and very true:

$$\Omega = \Omega(\text{baryon}) + \Omega(\text{CDM}) + \Omega(\Lambda)$$
$$= 0.04 + 0.23 + 0.73$$
$$= 1$$

As Mr Micawber would have said, 'result, happiness'. For obvious reasons, this package is known as 'ΛCDM' cosmology, and it is one of the great triumphs of science.

The next phase in developing our understanding of the Universe, a work still in progress, is to account for the origin of the kind of galaxies we see in the Universe within the framework of ΛCDM cosmology. But before we can do this, we need to take stock of the material world – the different kinds of galaxies we have to explain – since, unfortunately, this does not just consist of a neat division into disc galaxies and ellipticals.

The visible parts of spiral galaxies like the Milky Way form the classic two-part structure of a disc and a central nuclear bulge, although in some cases the bulge is very small. The spiral arms are the most visible characteristic of the disc, but the large quantity of dust and gas is just as important because it provides the raw material for the formation of the hot, young stars of the disc, known as Population I. The stars of the bulge and the globular clusters around a disc galaxy are older Population II stars. Spirals come with and without central bars, which may be temporary features that all spirals grow at some time. Most bright galaxies are spirals, and it is now accepted that all disc galaxies have black holes at their hearts, like the one at the centre of the Milky Way. The largest spirals may contain as many as 500 billion stars.

Disc galaxies without spiral arms (sometimes known, for historical reasons, as lenticular galaxies) still have the basic disc and bulge structure, but lack the dusty clouds. They are mostly made of Population II stars, and the inference is that they have used up all their star-forming material and settled into a quiet middle age. Distant lenticular galaxies seen at various angles can hardly be distinguished from ellipticals, although if their rotation can be measured by the Doppler effect that is a sure indication of their true nature.

Elliptical galaxies do not rotate as a whole, but the individual stars in ellipticals orbit around the centre of the galaxy. In nearby ellipticals that can be studied in detail it is possible to pick out streams of stars following different orbits oriented in many different directions, like the star streams in the Milky Way but on a grander scale. This variety of differently oriented star streams is what gives elliptical galaxies their overall shape, which is, strictly speaking, ellipsoidal, like a squashed or stretched sphere. They are dominated by old, Population II stars, and superficially look rather like the bulge of a disc galaxy without the disc. At least some ellipticals do contain dust, often in rings around the centre of the galaxy, but star formation is not going on in them in a major way at present. Although most bright galaxies are spirals, the largest galaxies are giant ellipticals which contain more than a trillion stars and are hundreds of kiloparsecs across. But the smallest galaxies in the Universe also seem to be ellipticals, containing only a few million stars and typically only a kiloparsec or so across. The smallest of these dwarf galaxies are comparable in size to the largest globular clusters, which is probably a clue to the origin of globular clusters. We can only see such tiny galaxies in our neighbourhood, where half of the couple of dozen nearest galaxies are dwarf ellipticals. It is very likely that most of the galaxies in the Universe are dwarfs like this, but we cannot see them at great distances.

Anything which cannot be described as an elliptical or as a disc galaxy is classed as an irregular galaxy. Irregulars usually contain a lot of gas and dust, in which very active star formation is going on. Because there is no well-defined structure like that of a spiral, this produces patches of star formation dotted around the galaxy, giving it an irregular, patchy appearance on photographs. The Magellanic Clouds, two small galaxies in the gravitational grip of the Milky Way, used to be classified as irregulars but have now been found to have an underlying barred spiral structure, difficult to see because of the patchy nature of the star formation. Some irregulars may be remnants or pieces of larger galaxies that have been disrupted tidally by close encounters with other galaxies. Such close encounters can be seen occurring across the Universe. In some cases, galaxies can be seen passing close by one another, being stretched and distorted by tidal forces; in other examples, galaxies are colliding with one another and may be in the process of merging – an important clue, as we shall see, to the origin of the kinds of galaxies we see around us.

Encounters between galaxies can also trigger massive bursts of star formation, which astronomers refer to, prosaically, as starbursts. There is no formal definition of a starburst galaxy, but it is one in which the rate at which stars are being formed is so great that all of the available gas and dust would have to be used up in a time much shorter than the age of the Universe. So they must be transient phenomena. In some starburst galaxies, stars are forming at a rate of hundreds of solar masses a year, about a hundred times faster than the rate of star formation in our own Galaxy. This would typically use up all the available material within about a hundred million years, less than 1 per cent of the age of the Universe.

Some starburst galaxies, especially the smaller ones, appear very blue, because the light from them is dominated by hot, young blue stars. These galaxies contain little dust, presumably the consequence of having recently been disturbed by an interaction or merger with another system, which stirred the dusty gas clouds up and triggered the burst of star formation which has depleted these reservoirs. Individual bursts of star formation occur within these galaxies in compact clusters of stars up to 20 light years (6 or 7 parsecs) across, a hundred million times brighter than our Sun. At the other end of the scale, some starburst galaxies are very large and very red, and are detected at infrared wavelengths using instruments carried into space on satellites. This is because they are enveloped in huge quantities of dust, which absorbs the light from the young stars inside the galaxy and reradiates it at infrared wavelengths. X-ray telescopes see right through the dust, and reveal that many of these large starburst galaxies have double cores of activity. This suggests that they are the result of two large galaxies merging. The double core is formed from the two black holes, one from each of the merging galaxies, which have not yet themselves merged. Starburst galaxies were found to be common, once astronomers had the technology to look for them and knew what they were looking for.

The presence of black holes at their hearts also explains why some galaxies show signs of violent activity in their nuclei, with outbursts flinging material away into space. Such objects were discovered piecemeal over the course of many decades, using different kinds of telescopes observing in different parts of the electromagnetic spectrum – visible light, radio, infrared, X-rays, and so on. As a result, many different names were given to these objects, which are now thought to be members of a single family. The generic name 'active galactic nucleus', usually shortened to AGN, therefore embraces a variety of such objects going under names such as Seyfert galaxies, N galaxies, BL Lac objects, radio galaxies, and quasars. It is now thought that these are all powered by the same kind of process, involving matter falling in (or on) to a supermassive black hole, with the differences being only those of degree, not of kind.

When material falls onto a black hole, the gravitational energy associated with it is released, being turned into energy of motion (kinetic energy) as the material speeds up. The same thing happens on a smaller scale if you drop something out of an upstairs window. The object falls downward at an increasing velocity as gravitational energy is converted into motion; then, when it hits the ground the kinetic energy is converted into heat, shared out among the molecules in the ground, which move a little faster as that patch of ground warms slightly. The 'Hot Spot' technology used in TV broadcasts of sporting events such as cricket matches makes use of this technology to show exactly where a ball has struck.

The particles of matter in the stuff falling into a black hole also collide with one another and get hot as they try to funnel into the hole, forming a swirling disc of hot material known as an accretion disc. The gravitational field of a black hole is so intense that a great deal of energy can be released in this way – up to 10 per cent of the mass energy, mc^2, of the infalling material. If the core black hole in a galaxy has a mass of only a hundred million times the mass of our Sun, roughly 0.1 per cent of the mass of all the bright stars in the surrounding galaxy put together, then it would only need to swallow the equivalent of a couple of stars like the Sun each year to provide the energy output seen in the most active AGN.

All large galaxies probably go through a phase of such activity, settling down into quiet respectability, like the Milky Way, when all of the 'fuel' near to the central black hole has been swallowed. But they could be reactivated if an encounter with another galaxy shakes things up enough for a fresh supply of gas and dust, or even stars, to spiral into the black hole. Any stars that suffer this fate get ripped apart by tidal forces into their component particles long before they are swallowed.

The energy from the central source is often beamed out in two directions on opposite sides of the galaxy. This is probably because the accretion disc of material around the black hole prevents energy escaping along the 'equator'. Both matter and energy can be ejected from the central region of the galaxy as a result, sometimes forming thin jets which interact with their surroundings to produce lobes of radio noise on either side of the galaxy. The most active AGN, the class known as quasars, are so bright that it is very difficult – sometimes impossible – to see the stars of the surrounding galaxy in their glare. As a result, they look like stars in ordinary photographs and their true nature is only revealed by measuring their redshifts. They radiate typically as much as 10,000 times as much energy as all the stars in the Milky Way put together, and some can be seen, even using optical telescopes on the surface of the Earth, at distances greater than 13 billion light years, with redshifts greater than 6; many are known with redshifts bigger than 4, corresponding to a distance of some 10 billion light years. But quasars are exceptionally bright, and are not necessarily typical of their surroundings; happily, large numbers of much fainter distant objects, relatively quiet galaxies even closer in time to the Big Bang, have been detected using the Hubble Space Telescope (HST) pushed to its limits.

The importance of studying objects at great distances across the Universe is that when we look at an object that is, say, 10 billion light years away, we see it by light which left it 10 billion years ago. This is the 'look back time', and it means that telescopes are in a sense time machines, showing us what the Universe was like when it was younger. The light from a distant galaxy is old, in the sense that it has been a long time on its journey; but the galaxy we see using that light is a young galaxy. Early studies of quasars showed that they were more common when the Universe was younger, just as you would expect if they are powered by accretion and fade away when they have swallowed all the available material. Historically, this was one of the clues that tilted the balance of evidence in favour of the Big Bang model and away from the Steady State idea. But the deepest observations made by the HST, corresponding to a look back time in excess of 13 billion years, tell us much more.

There is one further curiosity about all this that should be mentioned. For distant objects, because light has taken a long time on its journey to us, the Universe has expanded significantly while the light was on its way. So although a look back time of, say, 4.25 years implies that we are looking at an object that is 4.25 light years away, a look back time of 4.25 billion years implies that we are looking at an object that was in a sense 4.25 billion light years away when the light started its journey, but is now significantly farther away – in this case, getting on for twice as far. (It is even more complicated than this, since the distance light has to travel starts increasing as soon as it sets out on its journey, but this oversimplification will suffice to make the point.) This raises problems defining exactly what you mean by the 'present distance' to a remote galaxy, not least since nothing can travel faster than light so we have no way of measuring the 'present distance'. So like other astronomers I shall use look back time as the key indicator of how far away an object is, without trying to convert this into distances for anything outside our local region of the Universe. The 'distances' referred to earlier in this chapter should really be regarded as the equivalent of look back times.

Among the many advantages that photographic and electronic recording methods have over the human eye, the most fundamental is that the longer they look, the more they see. Human eyes essentially give us a real-time view of our surroundings, and allow us to see things – such as stars – that are brighter than a certain limit. If an object is too faint to see, once your eyes have adapted to the dark no amount of staring in its direction will make it visible. But the detectors attached to modern telescopes keep on adding up the light from faint sources as long as they are pointing at them. A longer exposure will reveal fainter objects than a short exposure does, as the photons (particles of light) from the source fall on the detector one by one and the total gradually grows. In the most extreme example so far of the application of this process, between 24 September 2003 and 16 January 2004 astronomers exposed the HST for a total of a million seconds to a tiny patch of sky in the constellation Fornax that looks completely black in ordinary photographs. The electronic image-gathering took place in 800 separate exposures, which were stored electronically then combined in a computer to give the equivalent of a single exposure more than eleven days long. The resulting image showed that this seemingly blank piece

18. The Hubble Ultra Deep Field

of sky is filled with galaxies, some of which are seen by light which left them when the Universe was less than 800 million years old, at a redshift of about 7.

The image is known as the Hubble Ultra Deep Field, or HUDF. The patch of sky on the image corresponds to just one thirteen-millionth of the area of the whole sky, no larger than a grain of sand held at arm's length, and has been described by the astronomers involved as equivalent to looking through a drinking straw 2.5 metres long. Yet this tiny patch of sky contains roughly 10,000 galaxies visible in the HUDF image. The ones which are of particular interest here are the faintest and reddest of these galaxies, with the largest look back time. The light from these particular objects trickled in to the detector on the HST at a rate of just one photon per minute.

Although the HUDF contains many normal galaxies, including spirals and ellipticals, these more distant objects have a variety of strange shapes, and some of them are clearly involved in interactions with one another. Some of the galaxies seem to be arranged like links on a bracelet, others are long and thin like toothpicks, and there is a variety of other peculiar shapes. At these early times in the history of the Universe, there were no spirals and no ellipticals – nothing resembling the kind of galaxies in our neighbourhood. Astronomers interpret this as evidence that they have captured a snapshot of the early stages of galaxy formation, before the galaxies settled down into the kinds of regular structures we see in the Universe at more recent times. When they are able to look back even farther in time with the next generation of telescopes, they expect to see nothing at all – the so-called 'dark age' between the time when radiation and matter decoupled, a few hundred thousand years after the Big Bang, and the time when the first galaxies formed, a few hundred million years after the Big Bang. Detecting nothing would, in this case, be a triumphant confirmation of a scientific theory. The oldest objects seen in the HUDF may themselves be at the edge of the dark age, about 400 million years after the Big Bang, at a redshift of about 12.

The most remarkable thing about these galaxies – perhaps we should call them proto-galaxies – is that they exist at all at such times. In much less than a billion years, the Universe had gone from being a sea of hot gas to a place where clumps of matter big enough to grow into the galaxies we see around us already existed, and were holding back, by gravity, matter that would otherwise have been spread ever thinner as the Universe expanded. This can only have happened if there were some kind of seeds on which galaxies could grow, cores with a strong enough gravitational influence to overcome the universal thinning. The identification of those cores with supermassive black holes proved the last link in a model of galaxy formation which explains how galaxies like the Milky Way came to be the way they are, and ultimately, since we are part of the Milky Way Galaxy, why we are here at all.

Chapter 7
The origin of galaxies

Before looking in detail at the explanation of how galaxies got to be the way they are, it makes sense to take stock of the way the Universe looks today, so that we have a clear idea of what it is we are trying to explain. I have already described the nature and appearance of individual galaxies, and mentioned the fact that most galaxies occur in clusters that are held together by gravity. But there is another layer of structure to the Universe, which provides important clues to the origin of galaxies. On the very largest scales, galaxies (strictly speaking, groups of galaxies and small clusters) line up in filaments that criss-cross the Universe and meet each other at intersections where there are large clusters of galaxies. Between the filaments there are darker regions where galaxies are rare. An analogy that is often made is with the view from space of a large, developed part of the world, such as Europe or North America, at night. Roads that cross the country are lit up by street lights and the lights of vehicles, and converge on brightly lit cities; between the roads, the countryside is dark. The key difference is that the distribution of galaxies in the Universe is three-dimensional, forming a foamy-looking structure as seen from Earth, revealed in the latest redshift surveys of the nearby Universe, out to a redshift of about 0.5. Unlike clusters and superclusters of galaxies, these filaments are not gravitationally bound together; extending the analogy with roads, they are simply the routes along which galaxies are moving as different clumps of matter tug on one another. But their existence does reveal how much matter is doing the tugging.

The overall pattern in the distribution of galaxies in three dimensions has now been studied in great detail by teams of astronomers who map the distribution of millions of galaxies on the sky, using redshifts to establish their distances. These observations of the relatively nearby Universe can be compared with the pattern of hot and cold spots seen in the microwave background radiation, imprinted at a redshift of 1,000, and also with computer simulations of how galaxies could grow in a variety of different model universes. The theoretical understanding of the way the Universe began expanding says that during the fireball stage, when baryonic matter and radiation were closely linked together, space was criss-crossed by sound waves with all wavelengths up to the limiting size, mentioned before, set by the speed of light. After decoupling, as we have seen, the radiation still carried an imprint of the pattern made by the sound waves, while the baryons settled down into clumps of matter held together by gravity. By applying statistical techniques to analyse the pattern of galaxies seen in the Universe around us, astronomers have now been able to detect the signature of these sound waves (so-called 'acoustic peaks') in the distribution of matter itself.

In 2005, two teams using different analyses both reported that statistical variations in the distribution of galaxies seen in large three-dimensional surveys show the imprint of these sound waves from the Big Bang. Observationally, everything fits together. But the computer simulations tell us that it would have been impossible for structures as large as the ones we see in the Universe today to have grown from the ripples present in the Big Bang fireball in the time available since the Big Bang, if the only thing pulling the baryons into clumps was their own gravity. The point is that, although the sound waves may have been large in the sense of having a long wavelength, they were also very shallow, merely ripples in the cosmic sea.

125 Mpc/h

19. The simulation of the distribution of matter in the expanding Universe described in the text. This closely matches the observed distribution of galaxies

The need for some extra gravitational influence should hardly come as a surprise, since I have already discussed the evidence for the presence of dark matter from the way individual galaxies rotate and the fact that clusters of galaxies are bound together gravitationally. But this is a completely different piece of evidence for the existence of dark matter, and the computer simulations are so sophisticated that they can tell us precisely how much Cold Dark Matter is needed to do the trick.

Such simulations track the behaviour of individual 'particles' moving under the influence of gravity in the expanding model universe. Each particle corresponds to about a billion times the mass of the Sun, and in the largest simulations to date ten billion particles are involved, moving in accordance with the known laws of physics. The simulations begin with the particles arranged statistically in the same way that we know matter was distributed at the time of decoupling, and then move forward in a series of steps taking account of the way the universe is expanding. The simulations can be chosen to include the effects of different kinds of cosmological constant, different amounts of dark matter, and different values for the curvature of spacetime. The process takes a lot of computer time. To obtain the simulation shown in Figure 19, a cluster of Unix computers using 812 processors with two terabytes of memory performing 4.2 trillion calculations per second ran for several weeks. Overall, the simulation produced a series of 64 snapshots of the model universe at different stages of its development, corresponding to different times since the Big Bang and culminating in the present day.

The results are clear; statistically, the simulation looks just like the real Universe, which is why I chose it. It represents the only class of such models that look like this. Starting from the kind of pattern of irregularities seen in the microwave background radiation, the kind of distribution of galaxies we see in the Universe today can only be produced in 13 billion years if the Universe is flat, there is six times more Cold Dark Matter than there is baryonic matter, and the cosmological constant contributes some 73 per cent of the mass density of the Universe. It is, of course, the highly successful ΛCDM model. The key to the formation of the observed structure is that as soon as baryonic matter decoupled from the radiation and was free to go its own way, in regions of the early Universe where there was already a slightly greater density of dark matter this pulled the nearby baryonic gas into the gravitational equivalent of potholes, where clouds of gas became dense enough to collapse and form galaxies and stars, distributed in a foamy pattern across the Universe. In the dark voids between the bright filaments there is still nearly the same density of both baryons and CDM, and it only needed a small (that is, shallow) ripple here and there to create the conditions needed to get the gas clouds to collapse. Changing the analogy from the road network mentioned earlier, the bright filaments can be thought of as rivers along which baryons flow.

This is the framework within which astronomers now believe they have a good understanding of the way in which individual galaxies have formed.

Just after decoupling, the baryonic material was still far too hot to collapse very much even in the presence of dark matter. But, crucially, the dark matter, being cold, began to collapse immediately in places where the density was a little higher than the average. Until about 20 million years after the Big Bang, corresponding to a redshift of about 100, the Universe was still pretty smooth, but the Cold Dark Matter particles were beginning to pull themselves into gravitationally bound clumps capable of holding matter back against the outward expansion of the Universe. Starting from the kind of ripples present in the background radiation, by a redshift of about 25 to 50 the dark matter would have formed clumps containing about the same mass as the Earth but about as big across as our Solar System. Most of the mass of such spherical clouds was concentrated near the centre, and the clouds formed in this way had a strong enough gravitational influence on each other to resist the universal expansion and form clusters, clusters of clusters, and so on in a hierarchical 'bottom–up' structure. This brought baryonic material streaming down onto the greatest concentrations of mass, forming stars and then galaxies at the nodes of the filaments as it did so and producing the filamentary 'cosmic highway' appearance of the Universe.

The first bright objects to appear in the Universe would have been massive stars, with between a few tens and a few hundred times as much mass as our Sun. These would have been very different from the stars around us today, because they contained only hydrogen and helium produced in the Big Bang, with none of the heavier elements. The first star-forming systems would have been part of a local filamentary structure which gradually became a subcomponent of a bigger filamentary structure stretching in a hierarchical fashion across the Universe, and still developing as clusters and superclusters of galaxies stream together in filaments. The models suggest that star-forming regions appeared about 200 million years after the Big Bang, each containing between a hundred thousand and a million times as much mass as the Sun, and between 30 and 100 light years across, similar in size to the clouds of gas and dust in which stars are forming today in the Milky Way. But these 'clouds' consisted chiefly of dark matter.

Simulations of the way baryons could coalesce to form stars in such clouds suggest that a filamentary structure like the large-scale filamentary structure developed inside each cloud, with matter concentrating at the nodes of the filaments. As the density increased, collisions between atoms became more common and some hydrogen atoms would have got together to make molecules of hydrogen. These molecules, crucially, would cool the gas inside the cloud by emitting infrared radiation; molecules of helium would do the same thing, but less efficiently. It was only this cooling that allowed the baryonic gas in the cloud to collapse still further into proto-stars, separating out the baryons to some extent from the dark matter.

In star-forming regions today, the cooling process is much more efficient thanks to the presence of heavier elements, which is why the clouds are able to collapse as much as they do before stars form. But in the primordial star-forming clouds everything happened at a higher temperature, with the result that the first star-forming knots in the cloud still had masses of a few hundred to a thousand solar masses. Just as in the case of stars forming today, it was very difficult for these clouds to fragment, and each cloud could have formed only a few (probably no more than three) stars, with some of the mass being blown away in winds as the proto-stars warmed up.

The result would have been a first population of stars (confusingly dubbed 'Population III' from an extension of the traditional nomenclature for stars in our Galaxy) with masses typically of a few hundred times the mass of the Sun and surface temperatures of about 100,000 K, radiating strongly in the ultraviolet part of the spectrum. This radiation, which filled the early Universe, is still visible today, but as a result of redshifting as an infrared glow detected by the Spitzer Space Telescope.

Although the first stars were bright, they would have been short-lived. The lifetime of a star depends inversely on its mass, because massive stars have to burn their fuel more vigorously to hold themselves up against their own weight. Within a few million years, still only about 200–250 million years after the Big Bang, stars which started out with masses roughly in the range from 100 to 250 times the mass of the Sun would have exploded completely at the end of their lives, spreading their material throughout the surrounding gas clouds. This material included the first heavy elements, which made cooling much more efficient when the next generation of stars formed, enabling the star-forming condensations in regions triggered into collapse by the blast waves from the exploding stars to become much smaller and make the first stars comparable to those in the Milky Way today. Indeed, some of those second generation stars may still be present in our Galaxy – the oldest Population II stars are calculated to have ages in excess of 13.2 billion years, so they formed within about 500 million years of the Big Bang.

Stars which have masses in excess of about 250 times that of the Sun are not completely disrupted in an explosive death. Instead, most of the material they contain collapses to make a black hole. These primordial stars formed in the densest concentrations of matter in the Universe at that time, so it is likely that the black holes would be close enough to each other for mergers to take place and the black holes to grow into even more massive objects. Nobody can be quite sure where the supermassive black holes at the hearts of galaxies today came from, but it seems at least possible that this merging of black holes left over from the first generation of stars began the process by which

20. A black hole at work. The jet emerging from the centre of the galaxy M87 is powered by a black hole. Although just visible in an optical photograph (left), the jet shows up much more clearly in the infrared (right)

supermassive black holes, feeding off the matter surrounding them, formed.

Observations of quasars at redshifts around 6.5 show that black holes with at least a billion times the mass of the Sun had formed long before the Universe was a billion years old. These are exceptionally large examples, which is why those quasars are bright enough to be seen at look back times close to 13 billion light years, but they confirm the speed with which galaxies appeared. Simulations show that there must have been many lesser black holes as well, forming the cores on which galaxies grew; each black hole may have been embedded in a halo containing a thousand billion solar masses of material. Baryonic material fell into the black hole, giving up gravitational energy to power quasars and other AGN, while stars formed in the quieter outer regions of what had become a galaxy as the baryonic material settled down; but the simulations also show that very large numbers of the original Earth-mass dark matter clouds should have survived all of this turmoil right up until the present day, and still be present in the dark matter haloes around the galaxies. It is estimated that there may be a thousand trillion (10^{15}) such objects in the halo of our Galaxy alone.

The calculations show that the process I have described can form an object as big as the Milky Way in the time available – a few billion years – provided that the central black hole has a mass of at least a million solar masses. Happily, observations reveal that the mass of the black hole at the heart of the Milky Way is about three million times the mass of our Sun. Everything fits. But although astronomers have a self-consistent model of how the first galaxies formed, there is still more to be explained, including an intriguing correlation between the mass of the black hole at the heart of a galaxy and the properties of the surrounding galaxy.

It is worth remembering how new the study of supermassive black holes of this kind is. They can only be studied directly in relatively nearby galaxies, where the presence of a massive central object is revealed by measuring the speeds of the stars orbiting near it, using the Doppler effect. The first supermassive black hole was only identified in 1984, and from then until the end of the 20th century simply finding one was an event; there were nowhere near enough of them known to make generalizations about their properties. But by the year 2000, the number of known supermassive black holes was up to 33, and one or two more are found each year. This is enough to begin to attempt an understanding of the relationship between such objects and their host galaxies.

At the beginning of the 21st century, astronomers discovered a relationship between the mass of the central black hole in a galaxy and the mass of the bulge of stars at the centre of the disk, or in the case of ellipticals the mass of the whole galaxy. There is no correlation with the properties of the disk itself; disks seem to have been added as an afterthought following the development of the bulge. Since the bulge at the centre of a disc galaxy closely resembles an elliptical galaxy, it seems likely that all primordial elliptical galaxies initially grew around black holes in the same way, but that not all of them then developed discs, possibly because of a lack of raw material from which a disc could form. So when referring to the generic properties of ellipticals and the bulges of disc galaxies, astronomers use the term 'spheroid'.

The masses of the supermassive black holes are determined by measuring the velocities of stars very close to the centre of a spheroid. The mass of the spheroid can be estimated from its brightness. But it is also possible to calculate the average speed of stars in the whole spheroid from an averaging of the Doppler effect for the larger system, providing a measure of what is called the velocity dispersion. This is a quite separate measurement, which can be used to reveal the mass of the spheroid in the same way that the motion of galaxies in a cluster reveals the mass of the cluster. Putting everything together shows that more massive black holes live in more massive spheroids. This is not really surprising. The surprise is that the correlation between the two masses is very precise – the central black hole always has a mass of 0.2 per cent of the mass of the spheroid.

This is such a tiny proportion of the total spheroid mass that it also demonstrates clearly that the black hole itself is not responsible for how fast the stars in the spheroid move; all that they 'notice', gravitationally speaking, is their own total mass (i.e. the combined mass of the stars and any remaining clouds of gas and dust between the stars). In essence the spheroid doesn't even know the black hole is there – if you took it away, the galaxy would look and behave exactly the same way.

Although the correlation is most simply expressed in terms of mass, the more significant aspect is that the stars in the spheroid around a more massive black hole move faster. This is an indication that the cloud of baryonic material from which they formed collapsed more within its dark matter halo during the process of galaxy formation. In other words, black holes grew bigger in systems which collapsed more, suggesting that the collapse feeds the black hole as it grows. Black hole masses are determined by the collapse process. It seems very unlikely that supermassive black holes formed first and then galaxies grew around them; they must have formed together, in a process sometimes referred to as co-evolution, from the seeds provided by the original black holes of a few hundred solar masses and the raw materials of the dense clouds of baryons in the knots in the filamentary structure.

The details of how the this symbiotic co-evolution occurred have yet to be unravelled, but it is easy to see in general how the energy pouring out from a growing black hole will first influence the way in which stars form in the surrounding material and then shut off the growth and activity of the black hole at some critical juncture by pushing the surrounding clouds of gas and dust away, simultaneously shutting down the early phase of rapid star formation. This fits observations of starburst galaxies in which winds carrying as much as a thousand solar masses of material a year are seen flowing out from the central regions; such winds, while they last, will trigger star formation in dense interstellar clouds, which they squeeze as they blow upon them. While 0.2 per cent of the available mass gets swallowed by the black hole, about 10 per cent of the baryonic material gets turned into stars.

This relationship between central black hole mass and velocity dispersion holds for a range of black hole masses at least from a few million to a few billion times the mass of the Sun – across a factor of a thousand (three orders of magnitude). It also holds across the Universe from the present day out to at least a redshift of 3.3, when the Universe was only two billion years old. When this relationship was first discovered, it seemed that flat disc galaxies without a central bulge did not have central black holes either, but in 2003 astronomers discovered a black hole with a mass of between 10,000 and 100,000 times the mass of our Sun in the bulge-less disc galaxy NGC 4395. This is still supermassive compared with the Sun, but a flyweight compared with the kind of objects I have been describing so far. But although this galaxy doesn't have a bulge, there is a central concentration of stars with a velocity dispersion that would imply a black hole mass of about 66,000 times that of the Sun. In other words, the velocity dispersion and mass match the relationship found in much larger systems. It may be that all disc and elliptical galaxies harbour central black holes, although irregulars may not.

The relationship also holds for our own Galaxy, the Milky Way, and its near neighbour M31, the Andromeda galaxy. The black hole at the heart of the Milky Way has a mass of only three million solar masses, and there is a small central bulge; the mass of the black hole at the heart of the Andromeda galaxy is 30 million solar masses, and there is a correspondingly larger central bulge. The overall relationship between the Milky Way and the Andromeda galaxy also offers a clue to what happened to galaxies after they had formed along with their central black holes in the early Universe.

The processes I have described so far explain the origin of the smaller elliptical galaxies and the disc galaxies. But the giant ellipticals seem to have formed, as I have already hinted, through mergers of smaller galaxies. At present, the Milky Way and the Andromeda galaxy are moving together at a closing speed of hundreds of km per second. The two galaxies are not destined for a head-on collision, but within at most about 10 billion years they will have merged together to make one giant elliptical. There is some evidence that the Andromeda galaxy has grown to its present size by swallowing a moderately large companion, since it seems to have a double-core, but the anticipated merger between two full-blown disc galaxies will be much more spectacular.

As I have mentioned, stars are so far apart from one another, compared with their own diameters, that even if two galaxies do collide head-on there is very little chance of stellar collisions. The galaxies pass right through each other, with gravity distorting the shapes of the galaxies as it changes the orbits of their stars. Collisions do occur between giant clouds of gas and dust between the stars, and these clouds are also squeezed and distorted by gravitational effects, causing the waves of star formation seen in many starburst galaxies. Gas and dust ejected from each galaxy as they pass through each other will make streams of material within which new globular clusters may form. Then, the galaxies swing round one another and experience another interaction. The process continues, with the cores of the galaxies getting closer on each swing, until the two galaxies merge into a single system in which there is no obvious disc but a whole mass of stars within which there are streams moving at various orientations, some of them carrying a memory of the discs that used to be. The final merger of the two central black holes releases a blast of energy which triggers a final phase of starburst activity before the new giant elliptical settles down into a quiet life. The penultimate stage of such a merger can actually be seen in the galaxy NGC 6240, where there are two black holes a kiloparsec or so apart, moving together on a collision course at the heart of the galaxy.

It used to be thought that in the case of the Milky Way and the Andromeda galaxy the timescale for all this ran from about five billion years to about 10 billion years from now, after the Sun will have ended its life as a bright star. But in 2007 a team at the Harvard-Smithsonian Center for Astrophysics reported calculations which suggest that the distortion of the Milky Way could begin in only about two billion years, when there could conceivably be intelligent life in our Solar System to watch the events. Any such watchers would have to be extremely patient, though, since even on the basis of the revised timescale the merger will then take a further three billion years to complete. By that time, the ageing Sun would have been displaced into an orbit about 30 kiloparsecs from the centre of the merged system, roughly four times more than its present distance from the centre of the Milky Way. Although the jury is still out on whether this revised timescale should be accepted, the end result will be much the same, whenever it happens.

Close encounters can also cause galaxies to shrink. In rich clusters of galaxies the individual members (the 'bees' in the 'swarm') are moving so fast under the influence of gravity that they cannot merge, but sweep past each other in glancing encounters that strip dust and gas, and even stars, from each other, sending the material flowing out into intergalactic space, where it forms a hot fog which can be detected at X-ray wavelengths. The largest elliptical galaxies sit at the centres of such clusters, like a spider sitting at the heart of its web, devouring anything that comes near them and growing fatter as they do so.

About one in a hundred of the galaxies seen at low redshifts are actively involved in the late stages of mergers, but these processes take so little time, compared with the age of the Universe, that the statistics imply that about half of all the galaxies visible nearby are the result of mergers between similarly sized galaxies in the past seven or eight billion years. Disc galaxies like the Milky Way seem themselves to have been built up from smaller sub-units, starting out with the spheroid and adding bits and pieces as time passed. I have already mentioned the star streams that are interpreted as remnants of lesser objects captured by our Galaxy; another line of evidence that supports this idea, probing further into the past, comes from globular clusters, whose ages can be inferred with good accuracy by studying their composition using spectroscopy.

The first stars contained very little in the way of elements heavier than hydrogen and helium, while younger stars are enriched with elements manufactured inside previous generations of stars in a well-understood way. Each globular cluster is made up of stars with the same age, confirming that they formed together from a single cloud of gas and dust. But the clusters have different ages from each other, showing that they formed at different times. The oldest are a little over 13 billion years old, nicely matching our understanding of when the first galaxies formed. The spread of ages of the globular clusters supports the idea that the part of our Galaxy outside the original spheroidal bulge formed from hundreds of thousands of smaller gas clouds each with up to a million solar masses of material. Whenever a new gas cloud collided with the growing galaxy, it would send a shock wave rippling through the cloud and trigger a burst of star formation in its core, forming a new globular cluster. The bulk of the material from the cloud would be tugged by gravity and slowed by friction to become part of the growing disc of material around the spheroidal bulge. Some globular clusters would survive until the present day; others would get ripped apart by tidal forces if their orbits took them too deep in towards the centre of the Galaxy. But the computer simulations show that this whole settling down process only works on the timescale available, if at all, if there is dark matter contributing to the overall gravitational field – several times more dark matter than there is baryonic matter. Without dark matter, disc galaxies could not grow at all, and without dark matter there would not have been any spheroidal seeds for them to grow on in the first place.

Within this self-consistent framework, small irregular galaxies are simply seen as bits and pieces left over from the early days of the Universe. Although it is difficult to see smaller galaxies far away, it is possible to make allowances for this in interpreting the statistics; when this biasing is allowed for, the observations tell us that there were many more small galaxies when the Universe was young than we see around us today. This is exactly what we would expect if many of the small galaxies have either grown larger through mergers or been swallowed up by larger galaxies. At the other extreme, more than half of the mass of baryonic material in the Universe today has already been converted into giant elliptical galaxies, the largest of which contain several trillion (10^{12}) times as much mass as our Sun – equivalent to about ten galaxies like the Milky Way put together. These are seen as far back as a redshift of 1.5, but spectroscopic studies reveal that many of them were fairly old by then, and that the components from which they formed must have merged at redshifts of 4 or more. Nevertheless, although the great era of galaxy mergers may have happened more than 10 billion years ago, probably the most important point is that these processes are still going on today. Galaxies are still involved in interactions and mergers, and clusters are still forming into superclusters. In this sense, the Universe of galaxies is still young; it has yet to mature. But what will be the ultimate fate of galaxies?

Chapter 8
The fate of galaxies

The fate of galaxies depends upon the fate of the Universe. There are three basic scenarios to be considered, and although theorists have come up with many variations on the basic themes these subtleties do not significantly alter the three possibilities for the fate of galaxies. The first possibility is that the Universe will keep expanding in more or less the same way as today, with a steady acceleration. The statistics of observations available today favour this possibility, but not decisively enough to rule out the two other options. The second possibility is that the acceleration of the expansion rate will itself accelerate. The third is that the acceleration will switch over into a deceleration at some point in the not too distant future and the Universe will eventually collapse into a Big Crunch that is the time-reversed version of the Big Bang.

All these scenarios are speculative, and when we look at the timescales involved there is no point in talking in anything but round numbers, so we start with the present age of the Universe rounded off to 10 billion (10^{10}) years as a benchmark. Also, we know so little about the nature of dark matter that it is difficult even to speculate about what might happen to it in the distant future, so I shall concentrate on the fate of baryons, the familiar particles that we ourselves are made of.

If the expansion of the Universe continues for long enough, then eventually all of the available gas and dust will be used up, and star formation will cease. From studies of the history of star formation, revealed by observations of different populations of stars in nearby galaxies, and the rate at which stars are being formed today in our own Galaxy, astronomers infer that this will happen in about a trillion (10^{12}) years from now – when the Universe is a hundred times older than it is today. Individual galaxies will become redder and dimmer as their stars fade and cool, and galaxy clusters will be carried ever farther apart, making it impossible for any astronomers around at the time to look out into the Universe and see anything beyond their own cluster. As the stars within each galaxy die, they will settle into one of three states. Stars with masses similar to or less than that of the Sun will simply fade away into cinders called white dwarfs, lumps of star-stuff containing about as much mass as our Sun in a sphere that is as big as the Earth. Stars which end their days with slightly more mass than this will shrink even further, forming compact balls with the mass of the Sun squeezed into the volume of Mount Everest; such neutron stars contain material at the density of an atomic nucleus. If there is even more mass left over when a star dies, or if a neutron star accretes enough mass from its surroundings, it will collapse all the way into a black hole.

Galaxies also shrink, on these very long timescales. This is partly because they lose energy through gravitational radiation, which has a trivial effect on any human timescale but adds up over trillions of years. They also shrink because of encounters between stars in which one star gains energy and is ejected into intergalactic space, while the other loses energy and falls into a tighter orbit around the galactic centre. In the same way, clusters of galaxies will shrink, and eventually both individual galaxies and clusters of galaxies will fall into supermassive black holes built up by this process.

You could regard this as the end of the story, since nothing recognizable as a galaxy would still exist at that time. But there will still be black holes and baryons around, in the form of ejected stars and traces of gas. If there is enough time, then according to particle physics theory even these ultimate constituents of the Universe will disappear. To indicate the timescales involved, for the moment I shall ignore the cosmological constant, and look at the old picture of a universe expanding steadily but more slowly as time passes, giving us infinite time to consider.

Theory tells us that the same processes which made matter out of energy in the Big Bang would eventually turn matter into energy as the Universe ages. 'Eventually' is the key word. Atoms are made of three kinds of particle: electrons, protons, and neutrons. Electrons are truly fundamental, stable particles; but neutrons, left on their own outside an atom, will decay into a proton and an electron in a few minutes. Protons seem to be stable on timescales comparable to the present age of the Universe, but theory says that even protons decay eventually, each one turning into a positron (the antimatter equivalent to an electron) and energetic gamma rays. Something similar happens to neutrons in white dwarfs and neutron stars, in this case with each decay producing both an electron and a positron to keep the overall balance of electric charge. The equations that describe how matter was produced in the Big Bang suggest that in any lump of ordinary matter half of the protons will decay in about 10^{32} years. Turning this around, in a lump of matter containing 10^{32} protons, one will decay every year or so. This is about the number of protons in 500 tonnes of anything – including water, butter, or steel.

This is a mind-bogglingly long time. Even 10^{30} is 10 billion cubed – a thousand billion billion billion – and 10^{32} years is a hundred times longer than 10^{30} years. But by about 10^{33} years from now, if the steady expansion carries on that long, virtually all the baryons not already swallowed up by black holes will have been processed in this way into electrons, positrons, and energy.

Whenever an electron and a positron meet, they annihilate each other in a puff of gamma rays. So all of the leftover star-stuff eventually ends up as radiation.

What of the black holes? Curiously, they suffer the same fate. There is a profound connection between the description of a black hole in terms of the general theory of relativity and thermodynamics and quantum theory. The key to this is the principle that lies at the heart of quantum physics, known as quantum uncertainty. This tells us that there are certain pairs of properties in the quantum world which match up in such a way that it is impossible for both members of the pair to have a precisely defined value at the same time. This is nothing to do with the imperfections of our measuring equipment, but is a feature of the way the Universe works. One pair of such variables is energy and time. In the context of the fate of black holes what matters is that the energy/time uncertainty tells us that there is no such thing as truly 'empty' space. If you imagine a tiny little volume of empty space, you might think it contained zero energy. But quantum uncertainty tells us that it *might* contain a certain amount of energy, E, provided it only does so for less than a certain time t. The bigger E is, the smaller t must be. So a little bubble of energy can pop into existence, then vanish, without ever being detected. Since energy can be equated with mass, this means that a pair of particles, such as an electron and a positron, could pop into existence out of nothing at all, provided they promptly disappear again.

Suppose this happened right at the edge of a black hole. Even in the tiny time available, one member of the pair could be captured by the black hole, while the other one escapes. But the Universe has not gained something for nothing; some of the mass of the black hole has been used up in the process, and the hole shrinks by a minuscule amount. The resulting avalanche of particles away from the surface of a black hole gives it a well-defined temperature, which is where thermodynamics comes into the story. The way the effect works, small black holes are hotter, and will evaporate away to nothing at all, exploding in a burst of radiation at the point where the mass inside the hole is no longer enough to close itself off from the rest of the Universe. A black hole with the mass of the Sun would take 10^{66} years for this to occur, even if it never swallowed any outside matter along the way. A black hole with the mass of a galaxy will evaporate in 10^{99} years, and even a hole containing the mass of a supercluster of galaxies – the biggest ever likely to form – will be gone in 10^{117} years. That really is as far as we can push our speculations and still pretend we are talking about the fate of galaxies.

But what if there isn't time for all this to happen? If the cosmological constant really is constant, and the expansion of the Universe accelerates at a constant rate, everything beyond the Local Group of galaxies, of which our Milky Way Galaxy is a member, will be carried out of sight within a couple of hundred billion years. Space outside our local bubble will be expanding faster than light, and no signal from outside will ever be able to reach any observers in the Milky Way, or whatever the Milky Way has become. There will be, in effect, a shrinking cosmic horizon defining the limit of observations. The processes I have just described will still go on, both outside this bubble and within it, but for all practical purposes within about 10 times the present age of the Universe there will be nothing to see outside the fading island of stars represented by whatever kind of merged supergalaxy has formed from the components of the Local Group. This is today's 'best buy' in terms of astronomical prognostications. There are, though, more dramatic possibilities. What if the cosmological 'constant' isn't really constant at all?

The supernova studies set limits on how much the cosmological constant could have changed as the Universe has evolved, but they are not yet good enough to prove that it really has been constant ever since the Big Bang. Perhaps it should really be called the cosmological parameter, to allow for the possibility that it changes as time passes. This has encouraged some theorists to speculate about how a changing value for the dark energy density of the Universe would affect the stretching of space and the fate of galaxies. The first possibility, that the rate at which the expansion of the Universe is accelerating may itself be accelerating, completely changes our view of our place in the Universe, since it suggests that, far from living in an early stage of a universe destined for a long life, we may already be a third of the way from the Big Bang to the end of everything material. Even more intriguingly, this idea suggests that if intelligence survives in the universe, observers will be able to watch this ultimate destruction almost to the end. (I use the lower case for 'universe' here to highlight that these are speculations about a possible universe, not certainties that apply to our Universe. My personal view is that these are fantasies, but entertaining ones!)

This scenario is sometimes referred to as the Big Rip, for reasons that will soon become obvious. It starts from the assumption that the expansion of the universe is responsible for creating dark energy, while, as I have explained, dark energy makes the universe expand faster. More expansion implies more dark energy, which implies a faster expansion, which implies more dark energy, and so on. All this is consistent with the known laws of physics, but is not required by those laws. If the cosmological parameter stays as small as it is today, objects like the Sun and stars, and galaxies, have no trouble resisting the cosmic expansion for hundreds of billions of years, because their gravity overwhelms the effects of dark energy. But in the runaway Big Rip scenario there soon comes a time when the dark energy, acting as an ever more powerful kind of antigravity, overwhelms gravity, and even what we think of as solid objects get torn apart by the expansion. This is an example of exponential growth, and, even in the most extreme version of the Big Rip allowed by the observations, although the end will occur in a little over 20 billion years time, nothing particularly odd will happen to objects like galaxies until the last billion years or so.

At that time, dark energy will overpower the gravitational forces holding the Local Group of galaxies together; this happens about 20 billion years from now, 10 times sooner than this will happen if the cosmological constant really is constant. By then, the large elliptical galaxy formed by the merger between the Milky Way and the Andromeda galaxy will still exist in a recognizable form, and although the Sun will have been dead for well over 10 billion years, there could very well be intelligent beings living on other Earth-like planets orbiting Sun-like stars, able to watch what happens as the size of the cosmological parameter continues to increase; the cosmological 'horizon' at that time would still be at a distance of about 70 Megaparsecs.

From this point on, it makes sense to measure the passage of events not in terms of the time that has elapsed since the Big Bang, but in terms of the time left before the Big Rip. Some 60 million years before the end, the galaxy – and all galaxies – would begin to evaporate as the dark energy became strong enough to overcome the gravitational attraction between stars, but it would still be possible for a planetary system like the Solar System to wander through space intact. Just three months from the Big Rip, the gravitational bonds holding planets to their parent stars would be loosened, and even any civilization that had the technology for observers to survive that catastrophe would reach its end when their planet was ripped apart by the cosmic expansion, about half an hour before the end of matter. In the last fraction of a second, atoms and particles would be ripped into nothingness, leaving a flat and empty spacetime. Some extreme versions of this idea suggest that a new universe might be born out of this void, and that our own Universe may have come from such a void. But as far as galaxies are concerned, we can say that the end is due, if this scenario is correct, in about 20 billion years' time and about 60 million years before the Big Rip.

Suppose, though, that the cosmological parameter fades away as time passes. It could fade away to zero, which would give us back the image of an ever expanding universe, with decaying matter and evaporating black holes, with which I started this overview. But why stop there? The equations allow for the possibility that the parameter could become negative. That brings doomsday even closer, perhaps as close to us in the future as the Big Bang is to us in the past. But this would be a different kind of doomsday – not a Big Rip, but a Big Crunch, equivalent to the Big Bang run in reverse.

Once again, I'll use the most extreme version of the scenario consistent with our observations of the real Universe and the known laws of physics. Just as a positive amount of dark energy acts like antigravity and makes the universe expand faster, a negative amount of dark energy acts like gravity and pulls the universe together, possibly reversing the cosmic expansion. Observations made so far, combined with theoretical considerations, suggest a range of possibilities for this kind of decline in the value of the cosmic parameter, implying that the Big Crunch could happen in as little as 12 billion years from now or as far into the future as 40 billion years from now. As before, the events are best described in terms of the time left before the end, which can also be expressed in terms of the shrinking size of the observable part of the universe. Since everything shrinks in the same way, even outside our horizon, exactly the same processes will be going on everywhere at the same time. But this time, intelligent observers would not be around to witness the death throes.

When the universal expansion halts and then reverses, it affects everywhere in the universe at the same time, because it is space itself that is affected by the changing value of the cosmological parameter. But because of the finite time it takes light to travel through space, any observers around just after the reversal, wherever they may be in the universe, would not see a universe dominated by blueshifted galaxies. Light from nearby galaxies would be blueshifted, but the light from distant galaxies, which had spent most of its journey travelling through expanding space, would still be redshifted. A long-lived civilization would be able to keep records showing the spread of a 'blueshift horizon' outwards at the speed of light, until eventually blueshifts do indeed dominate.

As far as galaxies are concerned, the collapse of the universe will have little effect for billions of years. The processes of star formation and galactic mergers that I have described will carry on just as before, with clusters of galaxies falling towards each other and eventually merging, and galactic mergers becoming ever more common, but still without any major problems for life forms like us living on planets like the Earth. The threat to life actually comes from one of the feeblest features of our Universe today: the background radiation left over from the Big Bang.

This cosmic microwave background radiation is left over from the fireball in which the Universe was born. Between 300,000 and 400,000 years after the Big Bang, at the time of decoupling, it was as hot as the surface of a star is today, and it has cooled all the way down to a temperature of 2.7 K (−272.3 °C) as it has stretched to fill the space available. But when the space available shrinks, the radiation will be blueshifted and compressed, heating up in the exact reverse of the process that cooled it. At the time when clusters have started to merge and all galaxies are beginning to be involved in mergers, the universe will be about one hundredth of its present size and the temperature of the sky will be about 100 K, still not enough to be alarming. But within a few million years, the temperature of the background radiation will exceed the melting point of ice, 273 K, and there would be no more snow or ice anywhere in the universe. Life might still be possible, but as the temperature continues to increase it moves up past the boiling point of water, 373 K, and soon the whole sky begins to glow brighter and brighter as time passes.

Life becomes impossible, and galaxies are disrupted into a mess of stars, a couple of billion years before the Big Crunch. A little less than a million years away from the end all baryonic matter that is not safe inside stars 'decombines' into its electrically charged components. Now, matter and radiation are recoupled into an intimate embrace. This is exactly the reverse of the decoupling that occurred after the Big Bang, and it happens at exactly the same time, 300,000 to 400,000 years away from the end, as decoupling occurred away from the beginning. The difference is that stars, or at least their cores, can still survive within this fireball, until the universe reaches one millionth of its present size and the temperature exceeds 10 million degrees, comparable to the temperatures inside stars. Then, even the stellar cores dissolve into the fireball. Eventually, everything disappears into a singularity, like the singularity at the heart of a black hole, or the one from which the Universe was born.

Which leads to the intriguing speculation that our Universe may have been born in *exactly* this way, out of the collapse of a previous universe, or a previous phase of our own Universe, which may follow a repeating cycle of expansion, collapse, and bounce. None of that, though, is relevant to the fate of the galaxies we see in our Universe. In the Big Crunch scenario, galaxies as we know them would be disrupted beyond recognition by about a billion years from the end, perhaps only 11 billion years from now.

But both the Big Rip and Big Crunch scenarios are speculations, offered here primarily to show the limits of what could happen. As far as we can tell, it is not possible for the Universe to recollapse in less than about 12 billion years, and nor is the Big Rip going to tear galaxies apart within the next 20 billion years. Thirty years ago, there was almost exactly the same uncertainty, between 12 billion years and 20 billion years, in astronomers' estimates of the time that has elapsed since the Big Bang. This has now been pinned down to 13.7 billion years. That's progress.

Maybe we can hope for similar progress over the next 30 years concerning our understanding of the fate of the Universe.

The present best buy prognostication for the fate of galaxies, however, is that the cosmological constant really is constant, and that although as a result of the gradual acceleration in the expansion rate of the Universe a Slow Rip may eventually happen, it will be so far in the future that it is scarcely worth bothering about. On that picture, galaxies are safe for a few hundred billion years, more than 10 times the present age of the Universe, and there will be plenty of time for other intelligent observers to work out exactly how it will all end.

Glossary

accretion disc: A disc of material orbiting around a star, black hole, or other object, from which matter spirals inward to fall onto the central object.

black hole: Any object with a gravitational pull so powerful that the escape velocity exceeds the speed of light. Supermassive black holes are the seeds of galaxy formation.

Cepheid: A kind of variable star whose properties make it useful in calculating distances across the Milky Way and to nearby galaxies.

Cold Dark Matter: The dominant material component of the Universe, present in about the ratio 6:1 compared with everyday matter. The presence of CDM is revealed by its gravitational influence, but nobody knows exactly what it is.

cosmological constant: A number which indicates the amount of dark energy in the Universe.

dark energy: Invisible form of energy, also known as the lambda field, thought to fill the entire Universe and act as a kind of antigravity, increasing the rate at which the Universe expands.

disc galaxy: A system of hundreds of billions of stars, most of them in a flattened disc, where there may be a spiral structure. Our Milky Way is a disc galaxy.

Doppler effect: Shift in the lines of a spectrum (for example, of a star) towards the red end of the spectrum if it is moving away, and towards the blue if it is moving towards us.

elliptical galaxy: A large system of stars with no obvious internal structure, with an overall shape like that of an American football.

escape velocity: The minimum speed needed for an object to escape from the gravitational clutches of another object. The escape velocity from the surface of the Earth is 11.2 km per second.

extinction: The dimming of light from distant stars by dusty material along the line of sight.

Galaxy (capital 'G'): The galaxy in which we live; also known as the Milky Way.

galaxy (small 'g'): Any one of the hundreds of billions of islands of stars in the Universe.

globular cluster: A spherical ball of stars found in the outer regions of a galaxy like the Milky Way. A single globular cluster may contain millions of individual stars.

Hubble Constant: A number which specifies how fast the Universe is expanding today. The rate of expansion changes as time passes.

Lambda (Λ) field: *See* dark energy.

Milky Way: A band of light across the night sky made up of vast numbers of stars too distant to be seen individually with the naked eye. *Also see* Galaxy.

nova: The sudden brightening of a star which makes it look like a 'new' object in the sky.

nuclear fusion: The process of fusing light nuclei (in particular, those of hydrogen) into heavier nuclei (in particular, those of helium). This releases energy and keeps stars like the Sun shining.

parallax: The apparent movement of an object across the sky when seen from different positions.

principle of terrestrial mediocrity: The idea that we do not occupy a special place in the Universe and that our surroundings are typical of those of a star in a disc galaxy.

spectroscopy: Technique of analysing the light from stars or galaxies by spreading it out into a spectrum.

spiral galaxy: *See* disc galaxy.

supernova: Extreme brightening of certain kinds of star at the end of their lives, when the single star can shine for a short time as brightly as a whole galaxy of stars like the Sun.

Universe (**capital 'U'**): The totality of everything we can see or be influenced by – the 'real world'.

universe (**small 'u'**): Term used to refer to a theoretical model, based on calculations and/or observations of what the world we inhabit might be like.

Further reading

Richard Berendzen, Richard Hart, and Daniel Seeley, *Man Discovers the Galaxies* (Columbia UP, 1984).

Peter Coles, *Cosmology: A Very Short Introduction* (OUP, 2001).

Arthur Eddington, *The Expanding Universe* (CUP, 1933)

John Gribbin, *Space* (BBC Worldwide, 2001).

John Gribbin, *Science: A History* (Allen Lane, 2002).

Alan Guth, *The Inflationary Universe* (Cape, 1996).

K. Haramundanis ed. *Cecilia Pagne-Gapschkin: An Autobiography and Other Recollections* (Cup, 1984)

Michael Hoskin, 'The Great Debate', *Journal for the History of Astronomy*, 7 (1976), 169–82.

http://antwrp.gxfc.nasa.gov/apod/ (for the observations in Hawaii, Chapter 3).

Edwin Hubble, *The Realm of the Nebulae*, Dover, 1958 (repr. of 1936 edn).

Malcolm Longair, *Our Evolving Universe* (CUP, 1996).

Denis Overbye, *Lonely Hearts of the Cosmos* (HarperCollins, 1991).

Martin Rees, *Before the Beginning* (Simon & Schuster, 1997).

Michael Rowan-Robinson, *The Cosmological Distance Ladder* (Freeman, 1985).

Thomas Wright, *An Original Theory or New Hypothesis of the Universe* (Chapelle, 1750; facsimile edn, ed. Michael Hoskin, Macdonald, 1971).

Index

Galaxies

Index

DARWIN
A Very Short Introduction
Jonathan Howard

Darwin's theory of evolution, which implied that our
ancestors were apes, caused a furore in the scientific
world and beyond when *The Origin of Species* was
published in 1859. Arguments still rage about the
implications of his evolutionary theory, and scepticism
about the value of Darwin's contribution to knowledge is
widespread. In this analysis of Darwin's major insights and
arguments, Jonathan Howard reasserts the importance of
Darwin's work for the development of modern science and
culture.

'Jonathan Howard has produced an intellectual *tour de
force*, a classic in the genre of popular scientific
exposition which will still be read in fifty years' time'

Times Literary Supplement

www.oup.co.uk/isbn/0-19-285454-2

COSMOLOGY
A Very Short Introduction
Peter Coles

What happened in the Big Bang? How did galaxies form?
Is the universe accelerating? What is 'dark matter'? What
caused the ripples in the cosmic microwave background?

These are just some of the questions today's
cosmologists are trying to answer. This book is an
accesible and non-technical introduction to the history of
cosmology and the latest developments in the field. It is
the ideal starting point for anyone curious about the
universe and how it began.

'A delightful and accesible introduction to modern
cosmology'

Professor J. Silk, Oxford University

'a fast track through the history of our endlessly
fascinating Universe, from then to now'

J. D. Barrow, Cambridge University

www.oup.co.uk/isbn/0-19-285416-X

INTELLIGENCE
A Very Short Introduction
Ian J. Deary

Ian J. Deary takes readers with no knowledge about the science of human intelligence to a stage where they can make informed judgements about some of the key questions about human mental activities. He discusses different types of intelligence, and what we know about how genes and the environment combine to cause these differences; he addresses their biological basis, and whether intelligence declines or increases as we grow older. He charts the discoveries that psychologists have made about how and why we vary in important aspects of our thinking powers.

'There has been no short, up to date and accurate book on the science of intelligence for many years now. This is that missing book. Deary's informal, story-telling style will engage readers, but it does not in any way compromise the scientific seriousness of the book . . . excellent.'

Linda Gottfredson, University of Delaware

'Ian Deary is a world-class leader in research on intelligence and he has written a world-class introduction to the field . . . This is a marvellous introduction to an exciting area of research.'

Robert Plomin, University of London

www.oup.co.uk/isbn/0-19-289321-1